COLUMBIA GEOMORPHIC STUDIES

STREAM SCULPTURE ON THE ATLANTIC SLOPE

STREAM SCULPTURE ON THE ATLANTIC SLOPE

A STUDY IN THE EVOLUTION OF APPALACHIAN RIVERS

BY

DOUGLAS JOHNSON

PROFESSOR OF PHYSIOGRAPHY
COLUMBIA UNIVERSITY

NEW YORK
COLUMBIA UNIVERSITY PRESS
1931

Copyright 1931
COLUMBIA UNIVERSITY PRESS

Published December, 1931

PRINTED IN THE UNITED STATES OF AMERICA
THE PLIMPTON PRESS · NORWOOD · MASS.

TO
MY GRADUATE STUDENTS AND RESEARCH ASSISTANTS
IN RECOGNITION OF THE DEBT I OWE
TO THOSE DISCUSSIONS WHICH BOTH CLARIFIED
AND MUCH ENRICHED THE VIEWS RECORDED HERE

FOREWORD

The truth about the ridges and valleys of the Pennsylvania Appalachians appears, as it is traced farther and farther toward its entirety, to involve a more and more complicated succession of events. At the time of a walking trip in the summer of 1863, when I had my first sight of the water gaps by which the Susquehanna cuts through three even-crested ridges above Harrisburg, mention was still made with some respect of Rogers' daring comparison between the folds of the heavy Appalachian strata and the waves of a violently shaken carpet, and also of the belief so confidently expressed by Lesley that the great folds had been quickly worn down to their present forms, once for all, by the rush of a mighty oceanic flood. Twenty-five years later, when my acquaintance with the ridges and valleys was much more extended and intimate, the theory of the cycle of erosion had been only briefly outlined, the possibility that the Appalachians had suffered more than one such cycle was still hardly imagined, and the over-simple, three-fold genetic classification of rivers and valleys introduced by Powell was generally accepted, although it took no account of subsequents or obsequents. There was therefore something of the spirit of adventure in my suggestion in 1889 that the Appalachians had, after their folding, experienced a succession of erosional reductions to low relief, during which a progressive modification of an initially consequent drainage system had led to the gradual evolution of the adjusted drainage system which, with many long subsequent and short obsequent and resequent members, is seen in the rivers and valleys of Pennsylvania today.

My error at that time of attributing a Cretaceous date to the peneplain represented by the ridge crests and my failure to recognize a later and lower peneplain were soon detected and corrected; but the general acceptance of — or at least the lack of successful arguments against — the rest of my conclusions during a period of forty years lulled me into a satisfied conviction that my scheme of Appalachian drainage development was essentially true.

It requires a vigorous, keen, independent, and persistent mind to reinvestigate accepted conclusions, and it is precisely such a mind that Douglas Johnson possesses; for he has an altogether exceptional ability in reopening closed problems and in resettling them on sounder foundations than before. This was shown in one of his earliest studies nearly thirty years ago, when he examined a previously announced explanation for the course of the Tennessee River in the Southern Appalachians. A marked feature of his treatment of that problem was his pursuit of the argument behind the evidence that had satisfied other investigators, in order to make his own determination of its validity. A case in point involved the evidence for stream rearrangement in the Tennessee area, which was thought to be found in the distribution of the fresh-water mussels or *Unionidae*; evidence that had been accepted as compelling by other readers as well as by its proponents, although it was based on the tacit postulate that the distribution was accomplished only along water courses in association with the migration of divides. Johnson not only showed that mussel embryos may be carried over divides by waterfowl from one stream to another, thus completely invalidating the argument, but showed also that the actual distribution of *Unionidae* in various Appalachian streams is altogether inconsistent with the main postulate of the evidence. The new conclusions he then reached have not yet been set aside.

At a later time his explanation of a certain slightly emerged shore line as a result of the spontaneous interaction of shore forms and shore processes, wholly independent of any change in level of land or of sea, gave an entirely new interpretation of a long-closed problem. For example, the low wave-cut bench and bluff on the east coast of Florida had been taken by all observers as unquestionably indicative of a slight upheaval of the land; but Johnson showed that the bench merely records the level of effective wave action on the still-standing peninsula previous to the construction of the present offshore sand reefs.

It was through work of this keenly discriminating kind, the steady progress of which it has been a keen pleasure to watch, that Johnson's larger investigation of Appalachian evolution as presented in this book has been approached; an investigation which has linked together a whole series of problems, the solution of any one of which would be a notable achievement, while the logical enchainment of all of them is nothing less than a physiographic masterpiece. Among these problems, in the study of which Johnson has had the aid of some of his students for several years past, as is duly set forth, two deserve to be specified as particularly significant. One problem concerns the physiography of the fall line, as indicative of the attitude of a now vanished peneplain on which the strata of the Atlantic coastal plain may have extended far inland from their present retreating margin and, thus extinguishing the original consequent or northwestward drainage, have given rise to many superposed streams directed to the southeast. The other problem concerns a delicate variation in the height of ridge crest lines in relation to deep-cut water gaps and less deep wind gaps, as indicative of whether the gap-cutting streams gained their location by superposition from an inferred coastal plain cover before or after the present ridge crests were defined. The close scrutiny given to the facts that enter these problems and the ingenious argumentation employed in their interpretation are truly admirable. It is altogether improbable that a study so critically pursued can be far wrong in its conclusions. The process of superposition, which I had employed but timidly forty years ago, has evidently worked with greater effect and at an earlier date than was then allowed it.

But there is still another relation in which all the following pages should be carefully examined: that is, their relation to scientific method. Two essays of historic importance should here be recalled: one is Chamberlin's exposition of the value of multiple working hypotheses; the other is Gilbert's address on the inculcation of scientific method by example; for the principles set forth in both these essays are admirably exemplified in Johnson's discussion. He has given scrupulous care to the use of all the mental faculties employed in scientific analysis. Observable facts are first gathered in large number, and are critically compared, classed and generalized. Various alternative hypotheses explanatory of Appalachian evolution, each one hopefully representing a set of imagined conditions and processes in the past from which the observed facts of the present may reasonably follow, are then brought forward, and the consequences logically deducible from the assumed premises and the adopted postulates of each hypothesis are deliberately and ingeniously worked out.

Here it should be noted that Johnson's independence of convention is well illustrated by his invention of the heretical hypothesis that the Atlantic coastal plain originally had a far inland reach on a now vanished peneplain, as intimated above; that his keen power of analysis is exemplified in the thorough deduction of all the various and curious consequences that are inherent but latent in all his hypotheses, quoted or original; that his impartiality is exhibited in the equally hospitable treatment given to every hypothesis that he entertains and tests; and that his rigorous discipline is shown in the demand that a successful hypothesis shall do much more than explain only the facts it was invented to explain. This demand is met by marshaling the several sets of deduced consequences, one after the other, and confronting them with the appropriate facts; for in this way directed observation may quickly discover which set of consequences is best confirmed. Thus finally an unprejudiced judgment may be made as to which hypothesis involves previously unnoticed consequences which correspond best with the visible facts of today. Results thus reached are as well established as they can be at the present stage of physiographic inquiry.

Yet we should not have learned our lesson if these well supported results are taken as absolute finalities. Their value is only pragmatic, in the sense of presenting the best conclusions we can reach today. If the next half century sees as great progress in earth science as the last half century has witnessed — and we must all wish it may — new interpretations may be proposed and some of the results which we now accept may have to be modified or even abandoned. The probability is that modifications will be chiefly in regard to minor matters, and that the major results of today will stand.

For those who live to see it, the extension of the present discussion of Appalachian evolution over the whole length of that ancient mountain system — in so far as its northeastern part is not submerged in the Atlantic ocean and its southwestern part is not buried under the Gulf coastal plain — will undoubtedly be entertaining. And those who are not likely to live so long may at least entertain themselves by expressing the hope that Johnson himself will conduct that extension to as successful an issue as that of his Pennsylvania study, and also by making the prediction that, even if some of the results now gained may then be modified, the outcome of such extension will confirm the value of the method of investigation that is here so ably set forth.

WILLIAM MORRIS DAVIS

August 30, 1931

PREFACE

A new interpretation of the evolution of the Appalachian highlands and adjacent physiographic provinces is here presented.* This "theory of regional superposition" as it may be called, differs from current interpretations of the physical history of the eastern United States (1) in that it ascribes an important rôle to a vast erosion surface or peneplane of pre-Schooley (pre-Kittatinny) age extending over the Appalachian highlands, few if any vestiges of which survive today except under and along the margins of the Atlantic coastal plain; and further (2) in that it assumes a former extension of the coastal plain deposits far northwestward across the belt of Appalachian folds, the deposits being supposed to rest upon the pre-Schooley surface just mentioned.

The early streams of the Appalachian region, whether antecedent to the folded structures or consequent upon them, presumably flowed from southeast to northwest. Hence the present courses of the master streams, *toward* the southeast, have long presented a puzzling problem to geologists. In his classic monograph on *The Rivers and Valleys of Pennsylvania* Professor William Morris Davis attempted to solve the problem by a beautifully devised but highly complicated series of stream readjustments related to a succession of land movements. The theory here presented not only accounts for the southeast courses of the streams in a simple and rational manner, but offers a satisfactory explanation for a variety of features which, as Davis frankly stated, remained unexplained on the basis of the theory advanced by him.

The pages which follow first review some established principles of Appalachian geomorphology, and then briefly set

* The first five chapters of this volume were awarded the A. Cressy Morrison Prize in Natural Science for 1930 by the New York Academy of Sciences.

forth the theory of regional superposition with the aid of nine block diagrams showing successive stages in the physical history of the eastern United States. There are next presented a variety of considerations which seem strongly to support the new theory. Finally, attention is directed to certain far-reaching implications of the theory, and to a critical analysis of some current views on Appalachian history which must be superseded in case the theory be adopted. In connection with this part of the work there are offered for consideration new interpretations of the drainage evolution of eastern Pennsylvania and northern New Jersey.

While the conception of regional superposition of Appalachian drainage from a surface older than the Schooley peneplane is offered as a working hypothesis rather than as a demonstrated fact, the evidence in its favor is deemed sufficiently strong to warrant discussion of the theory and its consequences.

The study of which this volume is one product involves a detailed analysis of Appalachian topography by means of hundreds of projected profiles specially designed to reveal the significant elements of the landscape along contiguous belts of country one mile or two miles in breadth. Preparation of such profiles is a long and arduous task, one which has already consumed several years and must continue for several years more. This work is being accomplished with the aid of generous support from the research funds of Columbia University. The writer desires to express his appreciation of this support, and of the efficient help of all who coöperated in the work, especially of his former and present research assistants, the Misses Bray, Shields, Preuss, Blauvelt, Rom and Zernitz, and Dr. Raisz. Various aspects of the problem have been assigned to a number of graduate students in physiography, in some cases leading to the preparation of masters' essays or doctors' theses. Acknowledgments are due these students, both for their original contributions in the fields assigned to them and for the profit derived from discussing with them different phases of the general investigation. To Dr. Erwin J. Raisz I am specifically indebted for the skill with which he has transformed outline pencil sketches into the diagrams used to illustrate this paper. Acknowledgments are also due to William Morris Davis, Bailey Willis, Henry B. Kümmel, Lloyd W. Stephenson, Florence Bascom, Eleanora Bliss Knopf, Alfred C. Lane, N. H. Darton, A. K. Lobeck, and W. H. Twenhofel for criticisms and other courtesies during the preparation of the manuscript. A particularly heavy debt of gratitude is due Mrs. Wallace Knapp for her highly efficient services in caring for proof sheets and all other matters connected with progress of the book through the press, a labor in which she was ably assisted by the Misses Clara Rom, Dorothy Mugler, and Dorothy Wallace.

Douglas Johnson

August 1, 1931

CONTENTS

FOREWORD BY WILLIAM MORRIS DAVIS vii
PREFACE xiii

PART I
THE THEORY OF REGIONAL SUPERPOSITION

CHAPTER ONE
PREVIOUS THEORIES OF APPALACHIAN HISTORY . . . 3
   SOME ELEMENTS OF APPALACHIAN HISTORY . . . 5
      RECOGNITION OF TWO INTERSECTING PENEPLANES ON THE
      CRYSTALLINES 5
      RELATIONS OF PIEDMONT AND NEW ENGLAND UPLAND 11
      THE KITTATINNY-SCHOOLEY QUESTION 12

CHAPTER TWO
THE THEORY OF REGIONAL SUPERPOSITION OF
APPALACHIAN DRAINAGE 14

PART II
CONSIDERATIONS FAVORABLE TO THE
THEORY OF REGIONAL SUPERPOSITION

CHAPTER THREE
CONSIDERATIONS FAVORABLE TO THE THEORY OF
REGIONAL SUPERPOSITION 25
   PROBLEM OF THE LOWER CONNECTICUT RIVER . . . 25
   OTHER SOUTHEAST DRAINAGE 28
   DEGREE OF STREAM ADJUSTMENT 34
   DEGREE OF MODIFICATION OF CONSEQUENT DRAINAGE . . 35
   ALIGNMENT OF WATER GAPS AND WIND GAPS . . . 36
   APPARENT GREAT EXTENT OF THE FALL ZONE PENEPLANE 37

CONTENTS

SCARCITY OF COASTAL PLAIN REMNANTS INLAND FROM THE
FALL ZONE 39
LONGITUDINAL PROFILES OF THE PENNSYLVANIA RIDGES . 43

PART III
IMPLICATIONS OF THE THEORY OF REGIONAL SUPERPOSITION

CHAPTER FOUR
FORMER GREAT EXTENT OF THE ATLANTIC
COASTAL PLAIN 47
   SIGNIFICANCE OF THE FALL ZONE 52
   HISTORY OF NEW ENGLAND RIVERS . . . 53

CHAPTER FIVE
HISTORY OF PENNSYLVANIA DRAINAGE . . . 55
   GENERAL CONSIDERATIONS 55
   REVERSAL OF PENNSYLVANIA DRAINAGE . . . 59
   DISAPPEARANCE OF THE ANTHRACITE RIVER . . 63
   SUPERPOSITION OF THE SUSQUEHANNA ABOVE HARRISBURG 65
   DAVIS' DOUBTFUL CASES 69
   RELATION OF STREAMS TO THE NITTANY ARCH . . 73

CHAPTER SIX
THE RIVERS OF NORTHERN NEW JERSEY . . . 76
   LACK OF STREAM ADJUSTMENT ALONG THE WATCHUNG
   RIDGES 83
   UNADJUSTED STREAMS OF THE HIGHLANDS AREA . 89
   OFFSET GAPS OF THE SOUTHERN WATCHUNG CRESCENT . 90
   THE NORTHWEST-FLOWING STREAMS . . . 97

CHAPTER SEVEN
SUPERPOSED SUBSEQUENT DRAINAGE OF THE
WATCHUNG CRESCENT 103
   THE DAVISIAN INTERPRETATION 103
   THE ALTERNATIVE INTERPRETATION 108
      GREAT BREADTH OF THE GAPS 108
      RELATIVE LEVELS OF PATERSON AND MILLBURN GAPS . 110

CONTENTS

Oblique alignment of gaps near Paterson	110
Direction of flow of superposed subsequent drainage	111
Objections to the capture theory	114
Significance of minor water gaps and wind gaps	117
Test of the theory of superposed subsequent drainage	122
Possible southwestward extension of the master subsequent	127
Larger significance of the superposed drainage	130
Derivation of the present drainage	130

CHAPTER EIGHT

Conclusion	132
Index	135

ILLUSTRATIONS

1. Rejuvenated Appalachians in Post-Newark Time . . . 15
2. The Fall Zone Peneplane 15
3. Encroachment of Cretaceous Sea and Deposition of Coastal Plain Beds 15
4. Arching of Fall Zone Peneplane and Its Coastal Plain Cover. Regional Superposition of Southeastward-flowing Streams 17
5. The Schooley Peneplane 17
6. Arching of the Schooley Peneplane 17
7. Dissection of the Schooley Peneplane and Erosion of the Harrisburg Peneplane on Belts of Non-resistant Rock 19
8. Uplift and Dissection of the Harrisburg Peneplane and Erosion of the Somerville Peneplane on the Weakest Rock Belts 19
9. Uplift and Dissection of the Somerville Peneplane to Give Present Conditions 19
10. Superposition of the Lower Connecticut, According to Tarr (After Davis) 26
11. Character of Early Drainage of the Northern and Central Appalachians, According to the Theory of Regional Superposition 29
12. Diagram Illustrating Manner in Which Remnants of Infaulted Strips of Coastal Plain Could Be Preserved Northwest of the Fall Zone 41

ILLUSTRATIONS

13. Evolution of Eastern Pennsylvania Drainage, According to the Theory of Reversed Consequent Streams 61

14. Evolution of Eastern Pennsylvania Drainage, According to the Theory of Regional Superposition   64

15. Cross Section of Northern New Jersey Showing the Geomorphic Evolution, According to the Theory of Regional Superposition in Pre-Schooley Time . 77

16. Northern New Jersey and Adjacent Regions . . 81

17. Inferred Cuesta and Lowland Topography above the Watchung Trap Sheets in Schooley Time . . 93

18. Superposition and Adjustment of Watchung Drainage, According to Davis 105

19. Diagram Showing the Supposed Course of Drainage in the Watchung Region Previous to the Last Glacial Invasion, According to Salisbury . . . 107

20. Development of Paired Wind Gaps 119

21. Hypothetical Course of the Hudson River in Schooley Time, According to the Theory of Superposed Subsequent Drainage 123

PART I
THE THEORY OF REGIONAL SUPERPOSITION

CHAPTER ONE

PREVIOUS THEORIES OF APPALACHIAN HISTORY

No present study of Appalachian geomorphology can fail to rest heavily on the great body of facts and inferences set forth in the literature by many workers in this classic field of investigation. One who would, from a new point of view, examine the facts afresh, must feel his great indebtedness to those who have pointed the way for his particular studies. In suggesting a theory of evolution of Appalachian topography which differs in one or two essential points from those current in the literature, the writer acknowledges his debt to predecessors in this field whose works are too varied for all of them to receive specific mention in the brief review of earlier theories presented in the pages which follow.

So far as the writer is aware none of the many workers on Northern Appalachian geomorphology has considered the possibility that the transverse drainage of this region was superposed from a coastal plain extending far inland across the Appalachian ridges and resting on a peneplane older and higher than the highest surface (Schooley or Kittatinny) preserved in the present topography. It is true that more than one investigator has wondered where the pre-coastal plain surface would pass if projected inland. It is true that Davis[1] inferred a former moderate inland extension of the coastal plain, and he followed a suggestion of Tarr's[2] so far as to admit the possibility of a superposed origin for the lower portions of the Connecticut and Housatonic Rivers, and explained certain New Jersey drainage and the lower Susquehanna on the same basis. But he limited his restitution of the coastal plain cover to a belt from 25 to 40 miles broad in Connecticut and New Jersey, and possibly 80 miles broad in central Pennsylvania; and he imagined this extension of the coastal plain cover to rest upon the Schooley peneplane. In his earlier work Barrell[3] went a little farther in Pennsylvania by carrying the cover over the Kittatinny Mountain ridge instead of to the vicinity of the ridge as did Davis; while in New Jersey he likewise inferred a cover over the Kittatinny ridge,* although not over the crystalline Highlands farther southeast nor over that part of the Kittatinny ridge in New York. The extent of the coastal plain cover in New England is less clearly stated by Barrell, although he seems to have fixed a "Cretaceous shoreline" in northwestern Massachusetts and Vermont, over a hundred miles from the present shore.[4] We have no sufficiently clear statement as to Barrell's latest views regarding the relation of the coastal plain cover to the surface known as the Schooley peneplane. Apparently his earlier view was similar to that held by Davis — the coastal plain deposits rested on the fluvial "Kittatinny" or "Cretaceous" peneplane (the Schooley peneplane of Davis); but later he certainly inferred marine planation at various levels, interrupted more than once by wholesale fluvial erosion, possibly continued to the stage of planation.[5] Barrell follows Tarr and Davis in ascribing the southeast courses of the Connecticut and Housatonic Rivers to superposition, but apparently it was with marine invasions of post-Schooley date that were associated the marine deposits upon which these streams acquired the courses in question.[6]

If any one has derived Appalachian drainage by superposition from a coastal plain resting on a peneplane of pre-Schooley (pre-Kittatinny) date, the fact has escaped the writer's notice. In any case this theory has made no headway in the literature; neither have the facts which make it seem plausible, nor the

* Often called "Blue Ridge" by Barrell through confusion with "Blue Mountain," an alternative name for the Kittatinny ridge.

radical changes which the theory, if accepted, must introduce into our interpretations of later Appalachian history, received the attention they deserve. It seems pertinent, therefore, to outline the conception of Appalachian evolution which the writer and his students have been employing as a working hypothesis for some years;* to specify certain facts which seem to give a measure of support to this view; and to indicate briefly some of the consequences one must be prepared to accept in case one adopts it. If in the present volume the writer necessarily devotes much of his attention to the new interpretation being set forth, the reader must not from this fact infer that earlier theories of Appalachian evolution have definitely been discarded. He should rather, if the evidence and arguments seem to warrant, give the new theory hospitable welcome into the group of multiple hypotheses entitled to consideration in any effort to discover the true history of the Appalachians.

Some Elements of Appalachian History

Recognition of two intersecting peneplanes on the crystallines. — It has long been recognized that the Northern Appalachians were (1) formerly reduced to a peneplane of remarkably low relief, (2) then partially overlapped by an encroaching coastal plain, and (3) later uplifted, the coastal plain cover largely stripped away, and the underlying folded structures deeply trenched by the superposed drainage. For Davis the widespread peneplane of this history was the Schooley surface, and he set moderate limits to the extent of coastal plain encroachment. Barrell apparently first accepted the idea of a fluvial peneplane, then inclined toward the conception of wholesale marine peneplanation; but he finally developed the view that an early fluvial peneplane of wide extent was later profoundly modified, and much of it completely destroyed, by

---

* For a brief preliminary statement of the theory, see program of the 1928 meeting of the Geological Society of America, published in *Bull.*, G. S. A., Vol. 40, 132–133, 1929.

marine agencies; and he assumed repeated marine transgressions which in New England were much greater than those recognized by Davis. We need not here concern ourselves with the age or the origin of the peneplane in question. But it is essential to our discussion to recall that Barrell [7] for New England, and Darton [8] and others for the Piedmont, some years ago recognized that the beveled crystalline surface underlying the Cretaceous coastal plain deposits along the Atlantic border, is steeper and may be older than the adjacent upland peneplane * — a view later set forth more fully by Shaw.[9]

This last point is so vital to the thesis of the present volume that it deserves some further discussion. Barrell's conception of marine erosion of the New England and Piedmont crystalline uplands rests largely on the recognition of what he considered to be wave-cut terraces on both surfaces.[10] The terraces were believed to trend roughly parallel with the present shore line, the descent from one terrace level to another being accomplished by a pronounced slope. Should the inclined crystalline surface under the Atlantic coastal plain represent one of these inter-terrace slopes, it would under Barrell's interpretation be *younger* than the adjacent upland, and certain of the deductions set forth in the following pages would be invalidated. Barrell does not discuss this possibility, although he does make a clear distinction between the inter-terrace slopes and the inclined floor upon which rest the coastal plain deposits. Doubtless some of the considerations set forth in later paragraphs seemed to him, as they seem to the writer, sufficient ground for regarding the pre-coastal plain floor as wholly distinct, in nature and in origin, from the slopes connecting the supposed marine terraces.

Multiple projected profiles east-west, north-south, and north-

---

* For purposes of the present discussion I shall assume the essential correctness of the old view that there is, both in the Piedmont belt and in southern New England, a major upland peneplane level easily recognized in accordant summits of hilltops in each region. Should it be proved that this surface is terraced, or that it consists of remnants of peneplanes of many different dates as different writers have suggested, the particular thesis of the present volume will not thereby be affected.

west-southeast, across all of southern New England, and similar profiles thus far completed for a part only of the Piedmont, enable one to analyze more fully than did Barrell the character of the supposed terraces, and the intervening slopes. This study has tended to increase the writer's early doubt [11] as to the marine origin of these forms. The problem is of too wide a scope for discussion here, but among the grounds for doubt may be mentioned (1) the very faint inclination of the supposed ancient cliff lines, quite unlike any marine forms known to the writer, and sometimes requiring several miles for a descent of but a few hundred feet; (2) the great discrepancy between steepness of the supposed weathered sea cliffs and steepness of weathered hill slopes of similar age, the supposed sea cliffs often being far more faintly inclined, instead of steeper as one might expect from analogy with known marine forms; (3) the presence of the supposed sea cliffs back of outlying highlands at points which should have been protected from marine erosion; (4) their absence from Mount Desert, especially well situated to record them;[12] and (5) the presence of slopes similar to the supposed sea cliffs in form, but facing inland away from the ocean. It has seemed necessary to look to non-marine agencies for a satisfactory explanation of the terraces.

At the beginning of this study the writer was impressed with the fact that if the sedimentary cover in the Grand Canyon district had been removed at the period of the early monoclinal terraces described by Dutton [13] as descending eastward from the formerly high Nevada-land like a giant stairway, the crystalline floor of that region would then have presented features similar to the rudely terraced crystalline uplands of the east. Thus, on the basis of rough analogy, and without formulating any specific supporting tectonic theory, the writer has tentatively harbored the idea that the New England and Piedmont terraces might in part at least be the surface expression of deep-seated changes associated with broad uparching of the Schooley peneplane.[14] One would not ordinarily expect so vast an area of sedimentary rocks to be uplifted without some bending or breaking of the mass; and the absence of a sedimentary cover which would clearly reveal monoclinal warps in the New England-Piedmont areas should not blind us to the possibility that the warps may be there, even if detected with difficulty in the much dissected crystallines. So also in the region of closely folded sediments, where warping is equally difficult to detect, the difference in elevation between the Schooley and Kittatinny levels discussed in a later paragraph has tentatively been ascribed in part to monoclinal displacement of a single erosion surface.

Full discussion of the New England and Piedmont terraces will appear in a later paper. The subject is raised at this point merely to acknowledge the justice of a criticism offered by Professor Bailey Willis. This student of the tectonics of continent building correctly observed [15] that so long as the difference of slope between the upland peneplane and the crystalline floor under the coastal plain may be ascribed to warping or faulting, the validity of important arguments presented in the following pages must remain in doubt. This observation is all the more pertinent because the present writer has himself been willing to admit the possibility that other changes of slope on the uplands may be of tectonic origin; and it gains weight from the fact that Professor Willis has elaborated a theory of continental uplift which, if accepted, may adequately account for monoclinal surface warpings or step faulting associated with deep-seated shearing and giving terraces roughly parallel with the axis of uplift. It thus becomes incumbent upon the writer to state the grounds which cause him to regard the pre-coastal plain floor and the adjacent crystalline upland as two distinct peneplanes intersecting at a low angle, instead of two parts of a single warped or faulted peneplane.

The two-peneplane interpretation has already been set forth in the writings of Darton and Davis (for the Piedmont), Barrell (for New England), Shaw, Renner, and other students of the question. As a rule the only evidence for two separate peneplanes specifically noted by these writers is the difference in slope, although other grounds are sometimes implicit in the published discussions. Among my own reasons for accepting this interpretation are the following: (1) The change in slope which marks the transition from upland peneplane to pre-coastal plain floor is more sharply localized, more angular, than seems appropriate for a monoclinal warping of the upland surface. In form the "fall-line angle" thus accords better with the theory that it results from the intersection of two erosion planes than with the interpretation that it is due to faint monoclinal warping, whether or not associated with faulting. (2) The great linear extent of the fall-line angle seems to place it in a class by itself. No other feature on the upland surface which might reasonably be interpreted as the product of warping or step faulting has been traced across so vast an extent of territory. It continues around the southern end of the Appalachians and where studied by Shaw[16] in the Mississippi embayment is strikingly apparent. (3) The observed close approximation of the fall-line angle to the eroded inner margin of the Atlantic coastal plain is a necessary corollary of the theory of two intersecting peneplanes, but is difficult to explain on the theory of a single warped or faulted surface. Shaw[17] has properly stressed the improbable nature of the assumption that an axis of warping would for so great a distance coincide with the inner margin of an earlier-formed coastal plain wedge. On the other hand, I have found it equally difficult to imagine reasonable conditions under which erosion would strip the coastal plain cover from a previously warped combined mass (crystalline floor and coastal plain cover) to give the observed coincidence. Neither does it seem probable that deposition would lay down the inner edge of a coastal plain everywhere close to the axis of an earlier warp, especially when we have no reason to believe that such a warp would maintain a constant elevation with respect to sea level.

Faulting might, under proper conditions, determine the inner margin of the coastal plain; but one of the requisite conditions is significant displacement along the fault plane, and the fall line not only affords no evidence of such displacement, but does not appear to be genetically associated with any fault. (4) The indefinite southeastward descent of the relatively steep pre-coastal plain floor clearly differentiates it from the strictly limited descent of the inter-terrace slopes of the upland. The first is a broad regional feature, the apparently continuous steep descent of which has with the aid of well records been measured over a zone scores of miles in breadth; the latter are purely local features, each descending slope soon being cut short by the next lower terrace, usually within a space of from one to 3 miles, and always within less than 5 miles from its beginning. As seen in projected profiles the local terraces are situated upon, and are only minor departures from, the two major intersecting planes (pre-coastal plain surface and upland surface). (5) The degree of preservation of the upland peneplane seems incompatible with the theory that it is as ancient as the crystalline floor underlying the Cretaceous beds of the coastal plain. This point has been emphasized by Shaw [18] and more recently by Ashley.[19] It seems only reasonable to suppose that any land surface of Cretaceous age, continuously exposed to erosion since Cretaceous time, must have been long ago completely destroyed. Hence we must regard the present upland peneplane as distinct from, and of later date than, the peneplane underlying the Cretaceous sediments.

The evidence on the points noted above will be presented fully elsewhere. In the present outline of the theory of regional superposition of Appalachian drainage it is sufficient to warn the reader that if he be not convinced of the existence of two distinct peneplanes of widely different age intersecting along the fall line, he must remain equally skeptical of certain important conclusions (though not necessarily of all the conclusions) set forth in the following pages.

Relations of Piedmont and New England upland. — It is obvious that if the main upland peneplane of the Piedmont belt be of different date from the New England upland peneplane, as Davis and certain others have supposed, the observed intersection of a steep pre-coastal plain floor with a more gently sloping Piedmont surface near Baltimore and elsewhere, would throw no light on conditions in New England. It would still be permissible to regard the pre-coastal plain floor of the south as merely the seaward border of the Schooley (or Kittatinny) peneplane, not yet denuded of its overlapping coastal plain cover.

The New England area must tell its own story. Here we might expect to find (1) the Schooley peneplane arching down to pass beneath the coastal plain deposits, as Davis believed to be the case even after he accepted Darton's view for the Piedmont area; or we might expect (2) a post-Schooley peneplane, similar to the Piedmont, to intersect the steeper pre-coastal plain (Schooley) surface. A great number of projected profiles constructed in connection with this study abundantly confirm Barrell's conclusion that the pre-coastal plain surface is more steeply inclined and, as shown above, presumably older than is the New England upland peneplane. Other profiles drawn to connect the New England upland with the top of Schooley Mountain demonstrate with equal clarity the identity of these two surfaces, thus proving the correctness of Davis' idea that the New England upland is in fact an extension of the Schooley peneplane. Thus neither expectation based on supposed conditions in the Piedmont area is realized. We are forced to conclude that, in the New England area at least, the pre-coastal plain floor is the long-preserved remnant of a peneplane older than the Schooley. While this conclusion does not necessarily apply to the eastern border of the Piedmont, there is reason to believe that there also the pre-coastal plain floor is part of a pre-Schooley erosion surface, and it is so represented in our diagrams.

The Kittatinny-Schooley question. — Can it be that the precoastal plain floor is the Kittatinny peneplane of Willis,[20] which some regard as a distinct erosion surface older than the Schooley, partially preserved in occasional abnormally high erosion remnants in the Appalachians of Pennsylvania and elsewhere? Possibly, but the weight of evidence seems to be against such an interpretation. Full discussion of this problem must await completion of projected profiles for critical areas; but it has always seemed to the writer that the moderate differences in altitude of ridge crests in the Pennnsylvania-New Jersey region, often cited in support of distinguishing between a Schooley and a Kittatinny peneplane, may more reasonably be interpreted as due in part to warping or faulting of a single surface, and in part to the persistence on such a surface of low monadnocks where noses of pitching anticlines or synclines gave unusually broad areas of resistant rock. This view, shared by other students of the question, receives substantial confirmation in a recent study [21] of Appalachian wind gaps and water gaps made by Professor Karl Ver Steeg of Wooster College as part of the general problem of Appalachian evolution here discussed.

REFERENCES

[1] William Morris Davis: "The Triassic Formation of Connecticut," *U. S. Geol. Surv., 18th Ann. Rep.*, Pt. 2, 1–192, 1898, see p. 165; "The Rivers and Valleys of Pennsylvania," *Nat. Geog. Mag.* Vol. 1, 183–253, 1889; Geographical Essays, 413–484, Boston, 1909, see pp. 471–473.

[2] R. S. Tarr: Personal communication to William Morris Davis.

[3] Joseph Barrell: "The Piedmont Terraces of the Northern Appalachians," *Am. Jour. Sci.*, 4th ser., Vol. 49, 227–258, 327–362, 407–428, 1920, see p. 240.

[4] *Ibid.*, see pp. 418, 421.

[5] *Ibid.*, see pp. 230, 328, 410, 416, 423, 424.

[6] *Ibid.*, see pp. 424, 425.

[7] Joseph Barrell: "Central Connecticut in the Geologic Past," *Conn. Geol. and Nat. Hist. Surv.*, Bull. 23, 1–44, 1915, see p. 26.

[8] N. H. Darton: "Outline of Cenozoic History of a Portion of the Middle Atlantic Slope," *Jour. Geol.*, Vol. 2, 568–587, 1894, see pp. 570, 571. "Artesian Well Prospects in the Atlantic Coastal Plain Region," *U. S. Geol. Surv.*, Bull. 138, 1–232, 1896, plates VII and XIII. N. H. Darton and A. Keith: "Washington Folio," *U. S. Geol. Surv., Geologic Atlas,*

Folio 70, 1901, Economic Geology sheet. See also, William Morris Davis: "The Peneplain," *Am. Geol.*, Vol. 23, 207–239, 1899, especially p. 214 and Fig. 1.

[9] Eugene Wesley Shaw: "Ages of Peneplains of the Appalachian Province," *Geol. Soc. Amer., Bull.* 29, 575–586, 1918.

[10] Joseph Barrell: "Piedmont Terraces of the Northern Appalachians," *Amer. Jour. Sci.*, 4th ser., Vol. 49, 227–258, 327–362, 407–428, 1920.

[11] See *Bull. Amer. Geol. Soc.*, Vol. 24, 691, 1913.

[12] Erwin J. Raisz: "The Scenery of Mt. Desert Island: Its Origin and Development," *N. Y. Acad. Sci., Annals*, Vol. 31, 121–186, 1929, see pp. 138–139.

[13] Clarence E. Dutton: "Tertiary History of the Grand Canyon District," *U. S. Geol. Surv., Mon.* 2, 264 pp., atlas, 1882, see p. 115.

[14] Douglas Johnson: "Appalachian Studies I," *Bull. Geol. Soc. Amer.*, Vol. 40, 131–132, 1929 (abstract).

[15] Bailey Willis: Personal communication.

[16] Eugene Wesley Shaw: "The Pliocene History of Northern and Central Mississippi," *U. S. Geol. Surv.*, Prof. Paper 108, 125–162, 1918. Figs. 22 and 23.

[17] Eugene Wesley Shaw: "Age of Peneplains of the Appalachian Province," *Geol. Soc. Amer., Bull.* 29, 575–586, 1918, see p. 583.

[18] *Ibid.*, pp. 575–586.

[19] George H. Ashley: "Age of the Appalachian Peneplains," *Geol. Soc. Amer., Bull. 41*, 695–700, 1930.

[20] Bailey Willis: "The Northern Appalachians," *Nat. Geog. Mon.*, Vol. 1, 169–202, 1896, see p. 189.

[21] Karl Ver Steeg: "Wind Gaps and Water Gaps of the Northern Appalachians, Their Characteristics and Significance," *N. Y. Acad. Sci., Annals*, Vol. 32, 87–300, 1930.

CHAPTER TWO

THE THEORY OF REGIONAL SUPERPOSITION OF APPALACHIAN DRAINAGE

Whatever may be the final verdict on the Schooley-Kittatinny question, it is believed that the marginal, relatively steeply sloping, pre-coastal plain floor is part of a peneplane much older than Schooley or Kittatinny, which formerly projected far inland beveling the Appalachian folds. Presumably the observed remnants of this ancient surface, being marginal to the upwarp, slope more steeply than did portions farther inland; but it is inferred that in any case this early peneplane, if restored, would pass high above the hilltops of the present upland of central New England and the crests of the Pennsylvania ridges. In age this peneplane must be later than the infaulted masses of Newark beds (Fig. 1) which it is conceived to have beveled, and earlier than such portions of the Cretaceous beds as may be found resting upon it at any given locality. It may in part at least be Cretaceous in age, for contrary to the view expressed by some writers there is no inconsistency in supposing Cretaceous beds to lie upon a Cretaceous peneplane, provided only that peneplanation occurred in a portion of Cretaceous time earlier than that represented by the covering deposits. It may in part be Jurassic, or Jura-Cretaceous, or even late Trias-Jura-Cretaceous, in age. Because it is exposed to view in, and genetically related to, the zone characterized by falls and rapids in streams passing from the Appalachian oldland into the Coastal Plain province, it has been called the Fall Zone peneplane [1] (Fig. 2).

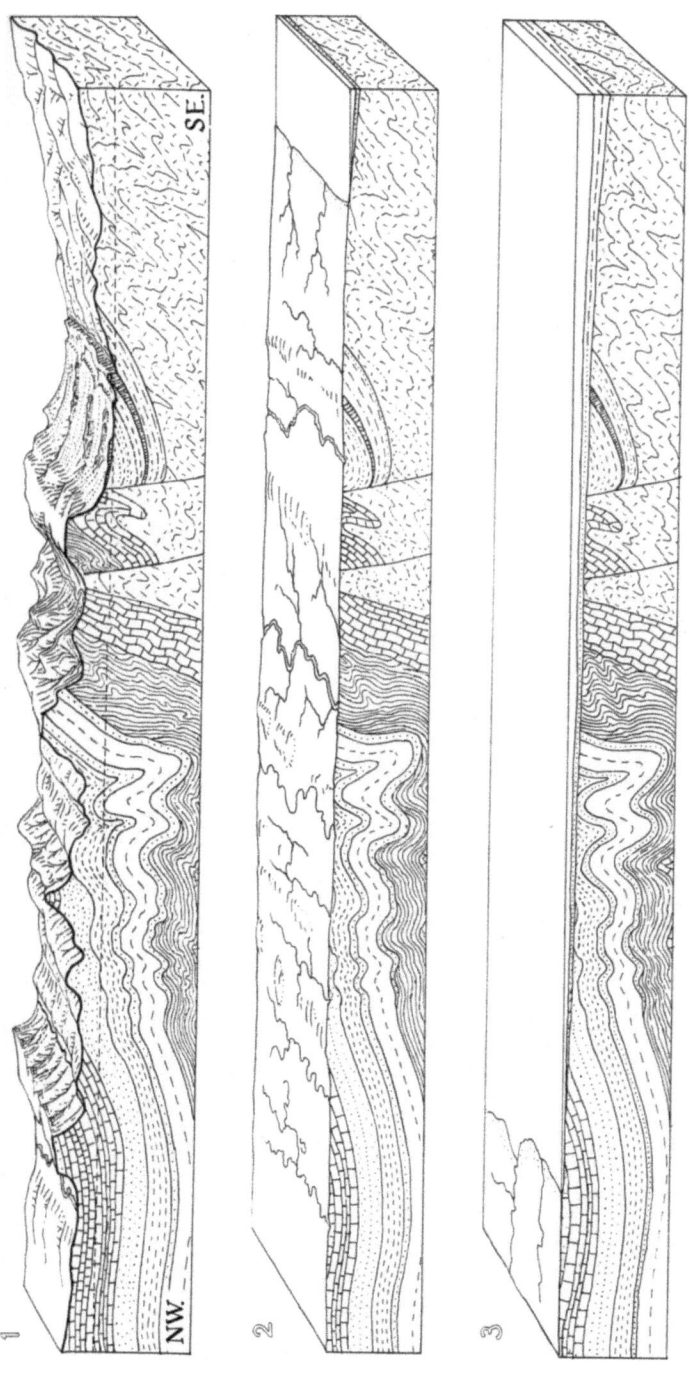

FIGURE 1. Rejuvenated Appalachians in Post-Newark Time

FIGURE 2. The Fall Zone Peneplane

FIGURE 3. Encroachment of Cretaceous Sea and Deposition of Coastal Plain Beds

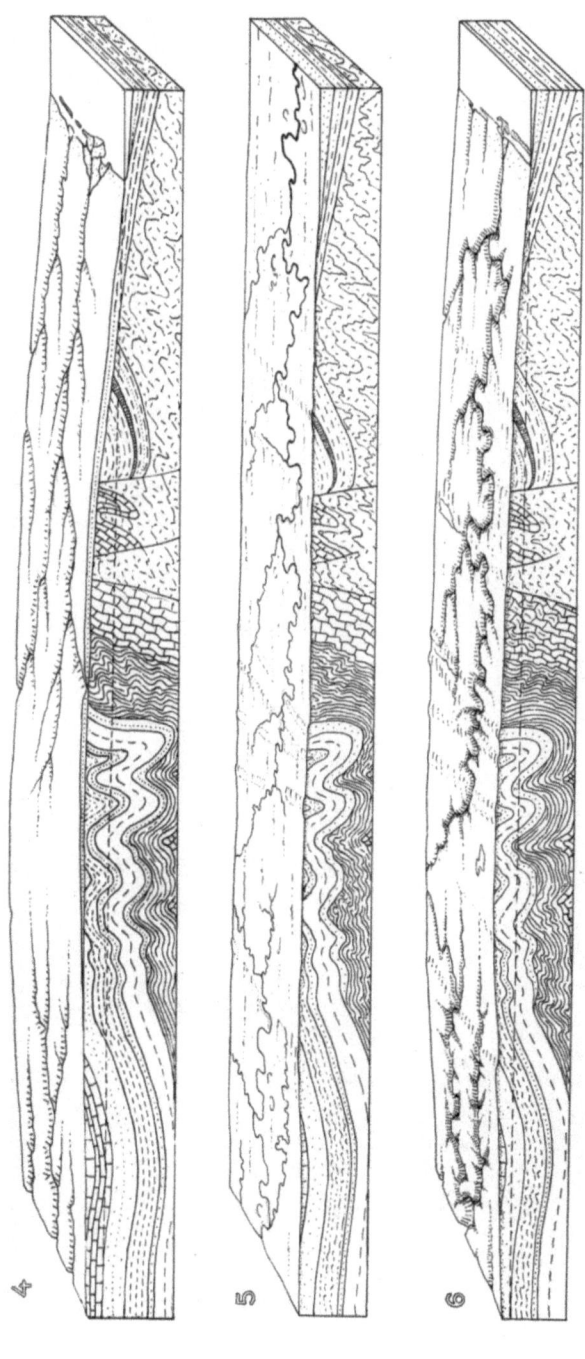

FIGURE 4. Arching of Fall Zone Peneplane and Its Coastal Plain Cover. Regional Superposition of Southeastward-flowing Streams

FIGURE 5. The Schooley Peneplane

FIGURE 6. Arching of the Schooley Peneplane

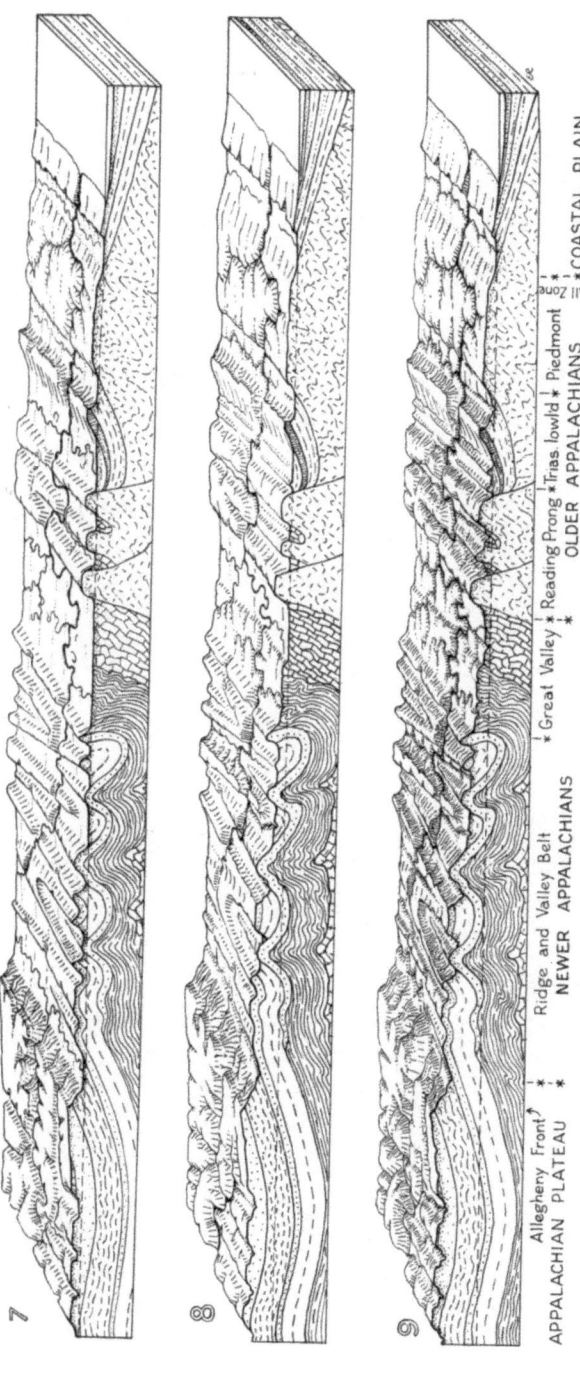

FIGURE 7. Dissection of the Schooley Peneplane and Erosion of the Harrisburg Peneplane on Belts of Nonresistant Rock

FIGURE 8. Uplift and Dissection of the Harrisburg Peneplane and Erosion of the Somerville Peneplane on the Weakest Rock Belts

FIGURE 9. Uplift and Dissection of the Somerville Peneplane to Give Present Conditions

Beginning in Cretaceous time (in some places perhaps as early as in late Jurassic time) the sea is supposed to have spread far inland over the Fall Zone peneplane, and to have buried that relatively even erosion surface and its continental deposits with a mantle of Cretaceous and possibly later sediments (Fig. 3). Thus all the drainage adjustments of the early cycle were completely obliterated. Later a broad uparching of the Appalachian region initiated upon the exposed covering deposits a new consequent drainage system (Fig. 4), traces of which may persist to the present day. Then during the Tertiary came the long Schooley cycle (or Kittatinny and Schooley cycles) of erosion, ending in production of the remarkably well developed and widespread Schooley peneplane (Fig. 5), which beveled both the ancient crystallines and the coastal plain deposits. Many superposed southeast-flowing consequent streams held their courses across the Appalachian structures, although during this long erosion period much adjustment of drainage to structure must have been accomplished.

Uparching of the Schooley peneplane (Fig. 6) permitted rejuvenation of the drainage and in due time development of the Harrisburg peneplane on weak rock areas (Fig. 7). Further moderate uplift allowed incision of streams in the Harrisburg level and development of the Somerville peneplane on rocks of excessive weakness, such as limestones and certain shales (Fig. 8). Finally, renewed uplift may account for the intrenchment of rejuvenated streams below the Somerville level (Fig. 9). Thus the later stages of Appalachian history are believed to follow the sequence set forth by Davis, Campbell, and other students of the region many years ago.

Attention is here centered on the major stages of Appalachian history, without discussing the existence of additional peneplanes reported by some observers. If intermediate peneplane levels exist in the present topography, these may readily be fitted into the scheme outlined above; while further study may show that the Schooley peneplane was not the first erosion surface to intersect the Fall Zone peneplane and bevel its coastal plain cover. Similarly, encroachments of the sea at other periods than the one pictured (Fig. 3) will fit into the story. It is not supposed that the oceanic waters necessarily everywhere advanced, once and only once, simultaneously to their farthest inland position, then uniformly withdrew to expose a coastal plain of simple structure and unvarying age succession. The single overlap pictured in Figure 3 may well be representative of a succession of advances and retreats, some of one age and some of another, some involving larger and others smaller areas, the whole giving a coastal plain cover, or a succession of coastal plain covers, of great complexity. Post-Schooley advances of the sea are equally admissible, provided only that such encroachments be not of such magnitude and duration as to require a substantial modification of the described sequence of events. Thus moderate marine transgression across the Schooley peneplane (Fig. 5), or over surfaces of later date, is permissible; but the overlapping of a coastal plain far across the Schooley surface in the manner pictured by Davis is definitely excluded, for reasons which will be made clear in a later paragraph. Finally, it must not be imagined that upwarpings of the continent were closely restricted to the periods of movement represented in the diagrams. Here again the true history was probably more complex than any drawings could effectively portray. We present certain principles in simplified form; but we do not lose sight of the greater complexity of Nature.

REFERENCES

[1] Henry S. Sharp: "The Fall Zone Peneplane," *Science*, N. S. Vol. 49, 544–545, 1929.

PART II
CONSIDERATIONS FAVORABLE TO THE THEORY OF REGIONAL SUPERPOSITION

CHAPTER THREE

CONSIDERATIONS FAVORABLE TO THE THEORY OF REGIONAL SUPERPOSITION

We have briefly presented a theory of Appalachian evolution through dissection by streams regionally superposed from a coastal plain cover resting on a peneplane of pre-Schooley (pre-Kittatinny) date. It is now proposed to set forth some of the considerations which appear to favor this theory, and which led to its adoption as a working hypothesis.

Problem of the Lower Connecticut River

Davis early called attention to the peculiar behavior of the lower Connecticut River, which just below Middletown in the state of Connecticut suddenly turns from its broad north-south valley cut in weak Triassic shales and sandstones, to pursue a southeast course through a narrow valley cut in resistant crystallines. As one possible explanation of this peculiarity he entertained the hypothesis that a large stream consequent on the Jurassic folding and faulting of the central Connecticut area may have found its exit from one of the Triassic basins southeastward along the course of the lower Connecticut.[1] But he seems to have inclined more strongly toward an interpretation [2] suggested to him by Tarr:[3] that the coastal plain sediments formerly lapped inland as far as Middletown, and that the lower Connecticut extended its course southeastward upon the coastal plain cover as the latter was gradually raised with a tilt in this direction (Fig. 10). From the coastal plain cover the stream was superposed upon the underlying crystallines; hence the present narrow valley in the lower portion of the river, and its apparently abnormal course. Kümmel[4] doubted whether the crystallines were worn low enough by the end of Jura-Cretaceous time to permit overlap of the coastal plain in Connecticut, and hence preferred Davis' earlier interpretation of the course of the lower Connecticut.

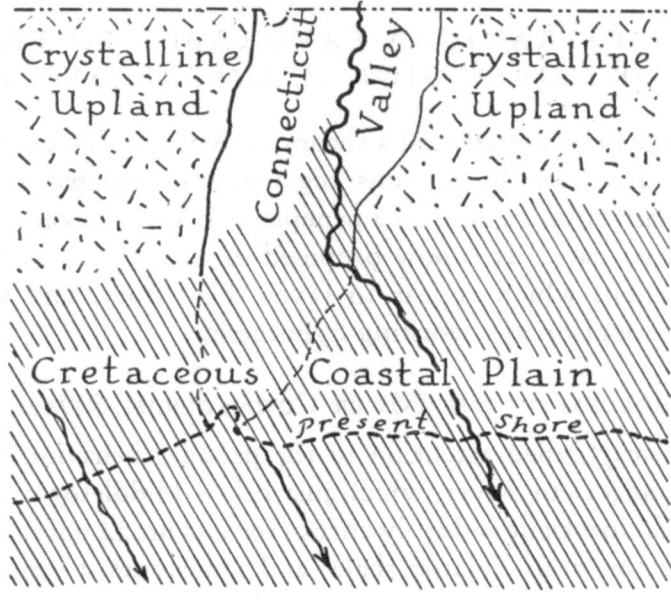

FIGURE 10

Superposition of the Lower Connecticut, According to Tarr
(After Davis)

On groups of projected profiles of southern New England a small but easily recognized angle between a less steeply sloping peneplane to the north and a more steeply sloping peneplane to the south can be traced across the state of Connecticut. This feature, early noticed by Davis and later discussed by Barrell, is interpreted as the angle formed by intersection of the Schooley peneplane with the Fall Zone peneplane. The two peneplanes are seen just over the words "Fall Zone" in the diagram, Figure 9, while their genetic relationships are made clear in the preceding diagrams of the series. It is at once evident that, on the interpretation here adopted, the angle marks the former inner margin of the coastal plain wedge which was left after Schooley peneplanation had beveled the Appalachians and their Cretaceous cover (Fig. 5). But the angle by no means indicates the maximum inland extension of the coastal plain; indeed its very presence seems to imply a much greater extension inland prior to the beveling effect of Schooley peneplanation.

If with these points in mind we reëxamine Tarr's explanation of the abnormal course of the lower Connecticut River, we find that it presents certain difficulties. In the first place we note the strong implication that the coastal plain must have extended much farther inland than the vicinity of Middletown. Accordingly, on Tarr's theory, we should expect the point of diversion toward the southeast to occur much farther north. If we substitute for the initial inner edge of the coastal plain the inner margin of the wedge left after Schooley planation (Fig. 5), we are not much better off. We must first doubt whether diversion from the weak Triassic belt across the coastal plain wedge is likely during the process of planation; and equally whether uplift of the completed peneplane, with a tilt to the southeast, would turn a master stream into a new southeastward course. In any case we face the further difficulty that the angle marking the former inner margin of the coastal plain wedge does not coincide with the point of deflection of the lower Connecticut. Instead the line marking this angle passes across the Connecticut Valley some miles south of the remarkable bend in the river's course.

Interpretation of the lower Connecticut as a remnant of early Jurassic consequent drainage presents difficulties no less serious. One might perhaps admit that the lower course of a large stream, even when so unfavorably situated as the lower Connecticut, could escape diversion in the long lapse of geologic time which witnessed at least two remarkably complete erosion cycles, the Jura-Cretaceous (Fall Zone) and the Tertiary (Schooley) epochs. But Davis[5] has presented grounds for believing that the Cretaceous sea advanced inland to the vicinity of Harrisburg in Pennsylvania, and to the crystalline highlands of northern New Jersey, blanketing with coastal plain deposits the areas thus encroached upon. In a preceding paragraph we have observed that in Connecticut the coastal plain cover not only reached the vicinity of the remarkable bend of the Connecticut River, but doubtless transgressed much farther inland. Acceptance of this apparently well supported conclusion involves the corollary that any Jurassic consequent drainage previously existing must have been extinguished by the Cretaceous marine transgression.

All of these difficulties disappear if we imagine the original Connecticut River to have been an initial consequent stream flowing southeastward across an uplifted coastal plain extending far inland over the Appalachian region (Fig. 4). On this interpretation the present lower Connecticut is a remnant of this consequent drainage, superposed on the underlying crystallines in late Cretaceous or early Tertiary time, and not diverted in the one complete cycle (Schooley) which has since intervened. The middle course of the river then represents a subsequent tributary which worked northward by headward erosion along the infaulted belt of weak Triassic rocks, and waxed great through conquest of other southeast drainage. The marked angle in the river's course should accordingly be in the Triassic belt near its contact with the crystallines, which is the case. There should be no relation between this angle and either the former innermost edge of the coastal plain, or the former inner margin of the wedge left after peneplanation; and this also seems to be the case.

Other Southeast Drainage

If the lower Connecticut were the only southeast stream requiring explanation, or if a series of streams flowing in other directions throughout their upper and middle courses suddenly turned southeast at a line reasonably to be interpreted as the

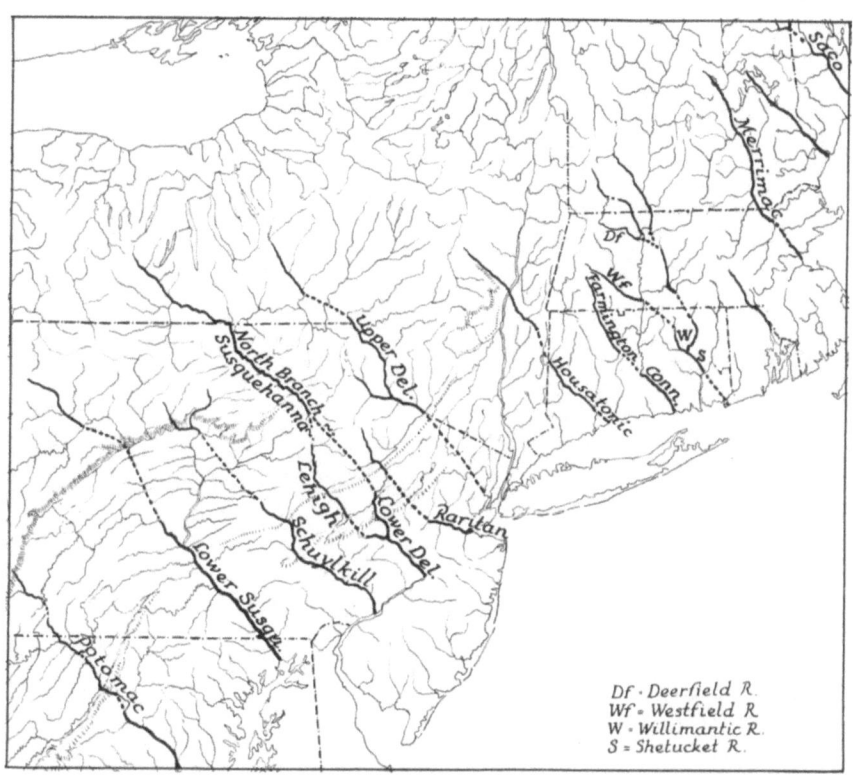

FIGURE 11. Character of Early Drainage of the Northern and Central Appalachians According to the Theory of Regional Superposition

Heavy solid lines show present southeast-flowing streams; heavy broken lines, assumed former southeast courses; light lines, other present drainage. The figure is chiefly designed to illustrate the *type* of drainage resulting from superposition, and no weight should be attached to specific stream connections except as indicated in the text. Each supposed former southeast stream course presents an independent problem, some of which have not as yet been adequately studied

inner margin of a former coastal plain, we might explain the phenomenon by invoking a moderate and comparatively recent marine transgression across both the coastal plain wedge and the crystallines shown near the right end of Figure 5 and subsequent diagrams. But the middle and upper courses of other streams show equally remarkable development of southeast drainage, oblique to the structures they traverse. The rectilinear southeast course of the middle Housatonic (Fig. 11) from the New York border to below Derby, Connecticut, and the course of the Hudson across the highlands might be accounted for on the above hypothesis without assuming any undue inland extension of such a coastal plain, although the *southwest* course of the lower Housatonic and certain other streams would still require explanation. But the southeast course of the Willimantic-Shetucket (Fig. 11, WS) would on this interpretation carry the old shore line far north only a short distance east of Middletown. And if we stop the marine transgression at Middletown to account for the thirty-mile southeast course of the lower Connecticut, how shall we explain the forty-mile southeast course of that upper portion of the Connecticut called the Farmington (Fig. 11), to say nothing of the southeast courses of the Westfield (Wf), the Deerfield (Df), and other New England rivers? Similar problems present themselves throughout the Appalachians to the southwest. Superposition of tributary streams from the flood plains of their masters, as suggested by Meyerhoff and Hubbell[6] for tributaries of the Connecticut, might locally be appealed to in certain cases; but both the prevalence of the southeast direction and the vast scale of the phenomenon seem to demand a different interpretation for the drainage as a whole.

The southeast course of the upper Farmington would, if prolonged, coincide with the southeast course of the lower Connecticut. Thus, with the exception of a fifteen-mile stretch in the weak rock belt of the Triassic Lowland (where destruction of the inherited course would most easily be accomplished;

where the deep, broad cut of Cooks Gap in the main trap ridge west of New Britain testifies to the former presence of a large stream; and where the Mattabesset River flowing southeast from the Gap reduces the actual interruption of southeast drainage to but five miles) we have a prominent continuous southeast drainage line from within Massachusetts clear across the state of Connecticut to the sea. Indeed, the most striking characteristic of New England drainage is its composite nature, large as well as small southeast-trending streams combining with the south- or southwest-trending streams to give a sort of latticed pattern.

In New York and Pennsylvania the North Branch of the Susquehanna (Fig. 11) pursues a remarkably direct southeast course for over a hundred miles; then, after a break of some fifty miles, a fifty-mile segment of the Delaware continues the southeast direction of the North Branch. Along this same line, and occupying the intervening space where southeast drainage at present has no great representative, is the remarkable wind gap near Pen Argyl, testifying to the former presence of a southeast-flowing stream of some magnitude. There is thus a strong suggestion that the North Branch of the Susquehanna formerly joined the Delaware near Easton to give a stream flowing directly southeast for more than two hundred miles. The suggestion is none the less strong if we grant that the present depth of the wind gap at Pen Argyl may have been cut by the beheaded remnant of the North Branch, after the main stream had been diverted westward to join the middle Susquehanna drainage. The lower Susquehanna (Fig. 11) and the lower Juniata together give over one hundred miles of southeast drainage, while the strikingly parallel southeast courses of the Schuylkill, the Lehigh, the upper Delaware from Broome County (New York) to Port Jervis, the Potomac from Hancock to Washington, and other streams, present a widespread phenomenon which any theory of drainage evolution must satisfactorily explain.

The remarkable development of southeast drainage in the Appalachian region has long challenged attention. So impressive are the rectilinear courses of many southeast-flowing streams that Hobbs[7] sought to account for them by the controlling influence of a system of fractures trending north 44 degrees west. He writes:

Examination of the rivers of the Atlantic border region, as we shall see, furnish evidence that dislocations approximating to the northwest-southeast direction have controlled.

Among the stream "lineaments" specifically mentioned by Hobbs are the lower Connecticut-Farmington, lower Delaware-upper Susquehanna, the lower Susquehanna, and the Potomac. Against the fracture-control theory must be urged the lack of sufficient evidence of major faults or other fractures coincident with the stream lines in question. We agree with Hobbs that the remarkable development of southeast drainage demands an explanation; but rectilinear stream courses cannot of themselves be accepted as evidence of dislocations, since such courses may result from a number of causes, among which superposition from an inclined coastal plain cover is one.

It is believed that the directness of the southeast courses, their great length, and their remarkable parallelism exclude the possibility that they are parts of an ancient northwest-flowing drainage, whether antecedent to the Appalachian folding or consequent upon it, reversed to a southeast direction by the Newark depression or by headwater growth of originally small Atlantic slope streams. That streams flowing northeast or southwest along the Appalachian structures should by tilting be turned into southeast courses on the Schooley surface we regard as impossible since the angle of tilt would have to be abnormally high in order to deflect a river from even so shallow a valley as it occupies on a peneplane. Deflection through capture of master streams by northeast-growing tributaries stimulated by a southeast tilt of the Schooley peneplane we consider highly improbable, since the observed angle of tilting is very low, and the obstacles to capture of master streams by tributaries are always appreciable, and especially so when capture must be effected across the structural belts of the country. Professor Bailey Willis [8] has suggested that a deep mantle of saprolite would facilitate the quick growth of southeast-flowing consequent streams as soon as the peneplane was tilted in that direction. It is difficult to determine, on purely theoretical grounds, how effectively a mantle of saprolite, and of alluvium derived from its erosion, might play the rôle of a coastal plain cover in bringing about the development and superposition of southeast-flowing master rivers; but our studies indicate that on the Appalachian peneplanes hard rock ridges persisted as linear monadnocks throughout the leveling process, that the resistant conglomerates and quartzites were far less susceptible to decomposing agents than were other rocks in the region, and that the slopes of the peneplane surfaces toward the streams and down the streams were much more pronounced than is commonly supposed. All of these factors seem unfavorable to the idea of rapidly developing southeast-flowing master streams on a saprolite-mantled, tilted peneplane. The simplest, and we believe the most probable, explanation lies in early superposition of an original southeast consequent drainage from an ancient coastal plain cover, followed by extensive development of subsequent streams along weak rock belts during the Schooley and later erosion cycles. In any case, regional superposition near the close of a cycle earlier than the Schooley seems clearly to be indicated.

Degree of Stream Adjustment

The extent to which subsequent drainage has taken possession of weak rock belts in the folded Appalachians is truly remarkable. The writer finds difficulty in ascribing to the usually inferred erosional history of the Appalachian region that degree of adjustment which obviously did obtain there at the close of the Schooley cycle. As Professor Davis [9] has suggested to me, the difficulty may in some measure be reduced if we assume that pre-Triassic erosion, which is known to have planed crystalline rocks in certain portions of the Appalachian province, likewise beveled the folded Paleozoics farther to the northwest. But in any case the opportunities for adjustment are enormously increased by making the Schooley a late instead of an early cycle. On the theory here advanced the folds were effectively beveled by the Fall Zone peneplane before the Schooley cycle was initiated. As the Cretaceous coastal plain cover was stripped from the earlier erosion surface the streams found weak rock belts advantageously exposed and inviting their attack. Under these conditions the Schooley cycle would easily suffice to render exceptionally perfect those adjustments for which the way had been prepared in preceding erosion periods.

The writer fully recognizes the fact that this is a matter not subject to demonstrable proof. Whether or not the degree of adjustment in a given region is too perfect for the work of any particular cycle or cycles must ever remain a matter of personal judgment. But while the foregoing paragraph can only be assigned the weight appropriate to an individual opinion, it seems necessary to take some account of the considerations there set forth.

Degree of Modification of Consequent Drainage

Had the southeast-flowing streams been superposed from a coastal plain overlapping the Schooley peneplane, it should be an easy matter to determine many details of the early drainage pattern. Not until the superposed streams had cut through the cover into Appalachian structures would there begin destruction of southeast-directed drainage through successive captures by growing subsequents. Thus the traces of the earlier drainage should be well engraved across transverse ridges, either in the form of wind gaps, or of water gaps occupied by beheaded and hence diminished rivers. Restoration of the original drainage pattern under such conditions would be a simple matter.

On the other hand, if superposition occurred on a surface earlier than the Schooley the case must be very different. Streams which persisted in their southeast courses through the Schooley cycle might remain to the present day, or leave unmistakable traces of their recent paths. But streams diverted in the Schooley cycle, and especially streams repeatedly dismembered and led off in different directions at different times, would ordinarily leave no evidence of their early courses which could be recognized after the region had, following the captures, been reduced to a new peneplane (the Schooley), uplifted, and maturely dissected in post-Schooley time with the accompaniment of still further drainage modifications.

Now it is precisely the combination of some southeast courses still followed or easily restored with the aid of gaps cut below the Schooley level, and other southeast courses reasonably inferred but difficult or impossible to locate with precision because the necessary evidence is wanting, which characterizes Appalachian drainage. We regard this as strong evidence of regional superposition from a surface earlier than the Schooley.

Alignment of Water Gaps and Wind Gaps

Water gaps of the present drainage, and wind gaps of recently diverted drainage, although cut in successive ridges of the Appalachians, are often definitely aligned in a northwest-southeast direction. As examples of this phenomenon we may cite the remarkable series of water gaps north of Harrisburg, cut by the Susquehanna through First or Blue Mountain and Second Mountain, and through Berry and Mahantango Mountains; the gaps of the Juniata through Tuscarora Mountain and Raccoon Ridge; the gaps of the Lehigh through Bear Mountain, Mahoning Hills and Blue Mountain; the gaps of the Schuylkill through Sharp Mountain, Second Mountain, and Blue Mountain; the four gaps of Shickshinny Creek and North Branch of the Susquehanna through the Pocono and Pottsville ridges near the west end of the Wyoming basin; and the three gaps of the Little Schuylkill through Nesquehoning, Pisgah and Mauch Chunk Mountains.

Davis recognized in this alignment of gaps an obstacle to the acceptance of his theory of drainage modifications in Pennsylvania. As he pointed out,[10] the normal dissection of folded mountains, and the modification of consequent drainage by subsequent streams, do not offer a reasonable explanation of gap alignment. The phenomenon might well be associated with antecedent rivers trenching folds gradually upraised across their fully established courses; but Davis realized that there are insuperable difficulties which prevent acceptance of the theory of antecedence. All difficulties vanish when we turn to the theory of regional superposition. By providing for southeast-flowing streams on a coastal plain cover the theory *requires* that when such streams are superposed on underlying Appalachian structures the resulting hard rock ridges must be cut by a succession of gaps aligned as the streams flowed — from northwest to southeast. Wherever the southeast drainage still persists, aligned water gaps bear witness to the fact of superposition. Where the southeast drainage has recently been deflected, aligned wind gaps afford evidence no less significant.

Apparent Great Extent of the Fall Zone Peneplane

From the Connecticut-Rhode Island region southwestward to Georgia we trace the Fall Zone peneplane, usually beveling much disturbed crystalline rocks which are either revealed by stripping of the coastal plain cover or encountered in well borings. Thence around the southern end of the Appalachians and into the Mississippi embayment we trace the same ancient surface beveling Paleozoic sediments. Throughout this great distance the perfection with which the ancient erosion plane was developed is abundantly indicated by the character of the stripped areas, by the relative simplicity of the contact between coastal plain deposits and older rocks, and by the fact that well borings through the coastal plain usually encounter the older formations at depths appropriate to the slope of the tilted pre-coastal plain floor. Nowhere in this long belt of country, so far as the writer knows, is there any indication that the coastal plain deposits were laid down on a mountainous topography of strong relief, although monadnocks on the peneplane, or some dissection of the peneplane prior to deposition of the Cretaceous beds, are not necessarily excluded.

It seems highly improbable that a well developed peneplane eroded for the most part on resistant crystallines should have extended for fifteen hundred miles in a northeast-southwest direction, without having extended great distances in other directions. We might, of course, assume that the present exposed strip of the old crystalline floor lay close to the northwest border of a widely developed peneplane, and hence that this particular peneplane never extended far inland beyond the present inner margin of the coastal plain. We should also consider the possibility that this erosion surface was in the nature of a piedmont sub-alluvial bench or rock pediment (later encroached upon by the sea), and hence bordered by mountains a short distance inland from the present fall zone. But while I have entertained both possibilities as working hypotheses, they seem to present greater difficulties than does the more reasonable assumption that we see at the fall zone merely a chance exposure of a pre-coastal plain floor which extended far in other directions than those revealed by chance. We may thus regard a great inland extension of the Fall Zone peneplane and its coastal plain cover, above the present position of the Schooley peneplane, as the most probable among several possibilities not subject to discrimination by conclusive critical evidence.

Scarcity of Coastal Plain Remnants Inland from the Fall Zone

On the hypothesis that the Cretaceous sea encroached upon the present upland peneplane, whether in the manner inferred by Davis or that supposed by Barrell, remnants of Cretaceous deposits might be expected to persist on its surface, since there remain portions of this surface which have not suffered profound dissection. It is, of course, conceivable that post-Schooley erosion could wash away the last vestige of the unconsolidated covering beds even where the upland surface was fairly well preserved on hard rock; but the completeness of such erosion has been a matter for some surprise and comment.

If, however, the Cretaceous beds rested on an older peneplane which passed inland high above the Schooley surface, it would be inevitable that Schooley erosion, cutting clear through the coastal plain cover and developing a peneplane in the hard rocks far below, must annihilate the last vestige of the uparched cover. Only southeast of the line (approximately the "fall line") along which the ancient peneplane and its cover dipped below the plane of Schooley erosion, or where post-Cretaceous faulting (Fig. 12) or folding had dropped parts of the Cretaceous cover abnormally low northwest of that line, could we expect to find remnants of Cretaceous beds preserved at the present time.

When we inquire into the facts we find that Cretaceous remnants are reported in places on the stripped portion of the Fall Zone peneplane, and more abundantly farther southeast of the line where this peneplane plunges below the level of the upland surface. On the other hand, northwest of this line Cretaceous deposits are rarely reported. Of the cases known to the author the Cretaceous age of some is doubted by competent authorities, who are inclined to assign the deposits to the late Tertiary or even to the Pleistocene. Such is the case with the beds near Lutherville north of Baltimore, mapped as Cretaceous by Darton [11] but later classed as Pleistocene by the Maryland Survey.[12] Other examples are associated with faults which may conceivably have dropped the beds abnormally low. Thus deposits in the Chester Valley-Plymouth Valley belt northwest of Philadelphia, classified as Cretaceous by Bascom [13] and others, occur as isolated remnants of what apparently was once a more continuous east-west strip resting unconformably on Paleozoic limestones. To the south the Paleozoic beds are in contact with pre-Cambrian crystallines along a great east-west fault, represented in the Philadelphia Folio as a thrust fault with downthrow to the north, where the Cretaceous remnants are found. The unusual position of these beds has excited comment. Some have doubted their Cretaceous age, but other specialists in coastal plain stratigraphy who have recently examined them are inclined to accept the earlier correlation. The low position of the beds has been attributed by Bascom to solution and removal of the underlying limestones. This would seem to imply more profound disturbance of the beds than is reported. If the structure be anything like that represented in the Folio, a satisfactory explanation of the facts might be found in renewed movement along the fault plane which previously had dropped the limestones low, or raised the adjacent crystallines high, or both. The history would then be essentially that represented in Figure 12, although in other similar cases of preserved infaulted strips of the coastal plain we might equally well infer normal instead of thrust faults, and the crystalline rocks might equally well appear on both sides of the fracture. If the contact between crystallines and limestones in the Chester Valley region be a thrust plane of very low angle, and one that has been corrugated by the latest Appalachian folding, as described by Knopf and Jonas,[14] renewed movement on this plane could hardly be invoked, although subsequent faulting, either steep-angle thrust or normal, would not be excluded.

Further study must be made of the innermost remnants of supposed or real Cretaceous beds and of their relation to the

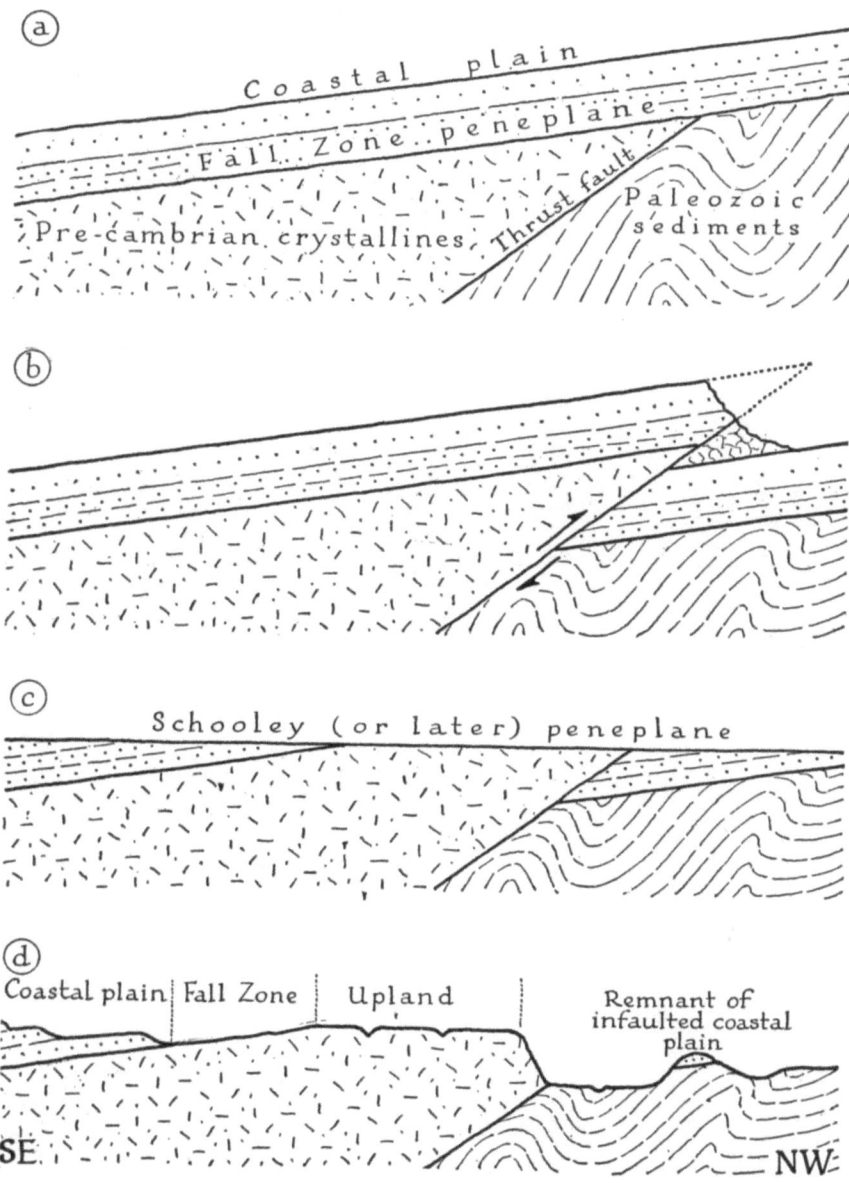

FIGURE 12

Diagram Illustrating Manner in Which Remnants of Infaulted Strips of Coastal Plain Could Be Preserved Northwest of the Fall Zone

angle between the two intersecting peneplanes as revealed by multiple projected profiles. Until such study is completed it is not possible to give a final verdict on the significance of the particular line of evidence here discussed. One may none the less, without impropriety, say this much: the undisputed facts thus far available favor the interpretation that the ancient Fall Zone peneplane with its coastal plain cover once arched inland above the present Schooley surface, and that the Schooley peneplane was developed later and therefore was not affected by the great marine transgression of the Cretaceous period. The conception of a drainage superposed before the Schooley cycle, and not after it, accords best with the history indicated.

Longitudinal Profiles of the Pennsylvania Ridges

What seems to be a striking confirmation of the theory of Appalachian evolution here outlined was discovered by Professor Karl Ver Steeg in his investigation of Appalachian wind gaps and water gaps.[15] Ver Steeg found that profiles of the ridge crests slope very faintly downward for long distances toward both the present water gaps and ancient water gaps changed by stream capture into wind gaps. The feature in question must not be confused with the " facets " of stronger inclination described by Barrell, and which Ver Steeg has also treated at length. The fainter slopes of ridge crests toward the gaps are accepted as evidence that the streams must have held their courses across the ridges throughout the Schooley cycle; for had the streams been superposed from either coastal plain or estuarine deposits overlapping the Schooley peneplane, the nice adjustment of streams to these faint depressions would be marred. The phenomenon is so widespread as to require a general explanation, and this is found in the idea of superposition from a coastal plain cover *before* instead of *after* the great Schooley denudation. It is perhaps pertinent to add that when the present writer suggested to Professor Ver Steeg that a study of Appalachian wind gaps and water gaps would make a useful contribution to Appalachian geomorphic history, it was not foreseen that such study would throw light upon pre-Schooley as well as upon post-Schooley events in the region. Thus Ver Steeg's discovery, being both independent and unexpected, may be considered as having special value in confirming the theory of Appalachian geomorphic evolution here set forth.

REFERENCES

[1] William Morris Davis: "The Triassic Formation of Connecticut," *U. S. Geol. Surv.*, 16th Ann. Rep., Pt. 2, pp. 1–192, 1898, see p. 155.

[2] *Ibid.*, see p. 165.

[3] R. S. Tarr: Personal communication to William Morris Davis.

[4] Henry B. Kümmel: "Some Rivers of Connecticut." *Jour. Geol.*, Vol. 1, p. 383, 1893.

[5] William Morris Davis: "Rivers and Valleys of Pennsylvania," *Nat. Geog. Mag.*, Vol. 1, 183–253, 1889. *Geographical Essays*, 413–484, Boston, 1909. "The Rivers of Northern New Jersey, with Notes on the Classification of Rivers in General," *Nat. Geog. Mag.*, Vol. 2, 81–110, 1890. *Geographical Essays*, 485–513, Boston, 1909.

[6] Howard A. Meyerhoff and Marion Hubbell: "The Erosional Landforms of Eastern and Central Vermont," *Sixteenth Biennial Report of State Geologist of Vermont*, 315–381, 1928.

[7] W. H. Hobbs: "Lineaments of the Atlantic Border Region," *Geol. Soc. Amer., Bull.* 15, 483–506, 1904.

[8] Bailey Willis: Personal communication.

[9] William Morris Davis: Personal communication.

[10] William Morris Davis: "Rivers and Valleys of Pennsylvania," *Nat. Geog. Mag.*, Vol. 1, 183–253, 1889, see pp. 252–253.

[11] N. H. Darton: "Geology of Baltimore and its Vicinity," Part II and accompanying geologic map. *Guidebook of Baltimore*, prepared for the American Institute of Mining Engineers, 125–139, 1892.

[12] E. B. Mathews: "Map of Baltimore County and Baltimore City, Showing the Geological Formations," *Md. Geol. Surv., Baltimore County Atlas*, 1925.

[13] F. Bascom, W. B. Clark, N. H. Darton, H. B. Kümmel, R. D. Salisbury, B. L. Miller, and G. N. Knapp: *U. S. Geol Surv., Geologic Atlas*, Philadelphia Folio, 1909. See pp. 8, 9.

[14] Eleanora Bliss Knopf and Anna I. Jonas: "Geology of the McCalls Ferry-Quarryville District, Pennsylvania," *U. S. Geol. Surv., Bull.* 799, 156 pp., 1929, see pp. 74–79.

[15] Karl Ver Steeg: "Wind Gaps and Water Gaps of the Northern Appalachians, Their Characteristics and Significance," *N. Y. Acad. Sci., Annals*, Vol. 32, 87–220, 1930.

PART III
IMPLICATIONS OF THE THEORY OF REGIONAL SUPERPOSITION

PART III

CHAPTER FOUR

FORMER GREAT EXTENT OF THE ATLANTIC COASTAL PLAIN

In the earlier pages of this volume we have outlined the "theory of regional superposition" of Appalachian drainage from a coastal plain cover resting on a peneplane of pre-Schooley (pre-Kittatinny) date, and have reviewed some of the considerations which seem to favor such an interpretation of Appalachian geomorphic history. In the pages which follow some implications of the theory, and some of its far-reaching consequences, are discussed.

If one accepts the theory of Appalachian evolution outlined in the first part of this discussion one must be prepared to admit that in the northern Appalachian region the Atlantic coastal plain formerly extended anywhere from 125 to 200 miles northwestward beyond the present limits of the remaining coastal plain deposits. This raises questions as to the location of land areas competent to supply material for the coastal plain, as well as to the existence in the Cretaceous or Tertiary series of deposits which could properly be interpreted as having been laid down far from the shore of a shallow epicontinental sea.

As regards the first point, the writer sees no inherent difficulty. The deposits resting directly upon the Fall Zone peneplane are not everywhere of the same age, and it seems quite possible that while the sea was encroaching upon certain portions of the beveled land mass other portions were still undergoing reduction and contributing significant amounts of sediment to the advancing waters. Large areas in the Appalachian interior may have remained high until long after those portions of the Fall Zone peneplane now observable had been brought low. The part of the extended coastal plain responsible for superposition of the southeast-flowing streams need not have been very thick, so the quantity of sediments needed to form it was not necessarily great.

The second point involves questions of more serious import. Many of those familiar with the Atlantic coastal plain deposits believe that they never extended more than a few miles northwest of their present inner margins. Stephenson, who has given some attention to the physiographic significance of coastal plain stratigraphy and paleontology, is inclined to limit the original northwestward extension of the formations to " 15 or 20 miles inland from their present bevelled outcropping edges."[1] This is only one-tenth of the extension implied in the theory of regional superposition of Appalachian streams presented in the first part of this volume.

Among the facts which are cited by those who accept the view of a very limited former extension of the Atlantic coastal plain toward the northwest we may quote the following: the occurrence of cross-bedding in certain of the formations, and the presence of cross-bedded layers interstratified with evenly-bedded sediments; the presence of large, thick-shelled molluscs, such as *Exogyra*, even in some of those deposits which seem of quiet water origin, this being taken to indicate that the deposits were laid down in relatively shallow, if quiet water; the presence of beds interpreted as delta deposits; the existence of fossil leaves, amber, lignite and lignitic sands and clays, in formations believed to represent near-shore deposition, perhaps partly in shallow marine waters and partly in lagoons or on low plains bordering the coast; the existence in the coastal plain series of numerous breaks, either unconformities or diastems, believed to indicate deposition at no great distance from shore where up and down movements of the land or changes in sea level are most readily reflected in the sedimentary record by such breaks.

Except for the larger unconformities, considered in a later paragraph, one may accept these facts as valid indications of shallow water conditions at certain times (in some cases even of terrestrial conditions), without finding in them evidence that the coastal plain never had a great extension far to the northwest.

Let us visualize the conditions as they are believed to have existed during the deposition of part, at least, of the coastal plain. The Atlantic border of the continent was a low-lying surface of faint relief, the Fall Zone peneplane. Over such a surface relatively slight changes of land level or of sea level would cause very extensive migrations of the shore line. At times the sea may have advanced far inland across the Fall Zone lowland; but it would be a shallow epicontinental sea, and sediments worked and reworked by waves and currents during advance and retreat of the shore would exhibit physical and faunal evidence of their shallow water origin. At other times the sea may have retired to the outer edge of the continental shelf, as suggested by Stephenson [2] for the Gulf coastal plain. During such retreat erosion might remove the exposed marine deposits, so that today there would remain within the present continental limits no trace of that part of the coastal plain series responsible for wholesale superposition of streams. Thus we must recognize that the theory presented in this paper does not call for the presence of deep water sediments and faunas in the coastal plain series; it does not even require the presence, in that part of the series available to inspection, of any of the sediments, whether of deep water or shallow water origin, upon which superposition of the Appalachian streams took place.

The theory does require that at some time during the coastal plain history there should occur at least one advance of a shallow sea across the low-lying Fall Zone peneplane to a line situated from 125 to 200 miles northwest of the present inner margin of the coastal plain province; or that continental deposits should combine with marine deposits to give a coastal plain cover of this great breadth. Formations of continental origin constitute part of the present coastal plain series, and may well have been of increasing importance farther northwest. On the other hand it seems doubtful whether, in view of the apparently remarkable parallelism of the original southeast-flowing streams, we can assign a preponderant rôle to deltaic or alluvial fan deposits. Such a possibility should not be excluded; but we believe the evidence favors a far northwestward advance of the sea. Extensive horizontal shifts of the shore line on low-lying lands is now recognized as a common event in geologic history. There is evidence that the eastern portion of North America has had its full share of such shifts, and Stephenson speaks of " the warping of the Gulf coastal plain in Upper Cretaceous time, with the accompanying transgressions and regressions of the sea, and the consequent forward, backward, and lateral shifting of marine faunas," all of which he regards as " analagous in certain respects to the broader continental warping of Paleozoic times which produced the ever-changing epicontinental seas, with their accompanying faunal migrations, so ably described by Ulrich, Schuchert, Butts, and other students of Paleozoic stratigraphy."[3] On the evidence of a notable unconformity between the Upper Cretaceous and the Lower Eocene beds of the Gulf coastal plain, Stephenson[4] infers one horizontal shift of the shore line in that region amounting to considerably more than 200 miles. This same unconformity, but recording a much greater time hiatus than in the Gulf region, is found in the coastal plain of New Jersey.[5] By analogy with the Gulf region we are thus led to infer a very great horizontal shift of the shore line in the New Jersey-Pennsylvania region. When we remember that the Atlantic coastal plain contains a number of unconformities, we are justified in concluding that a significant northwestward advance of the sea during one or more periods of coastal plain history is well within the realm of the possible.

In summing up his studies of glauconite in the New Jersey coastal plain deposits Mansfield[6] writes:

If, as seems probable, the conditions of the formation of glauconite in Cretaceous time were similar to those of today the depth of the sea water over New Jersey in the Navesink and Hornerstown epochs was 1,200 to 1,800 feet.

Mansfield appears to have derived these large figures from a misreading of Murray and Renard's somewhat ambiguous statement on page 383 of the *Challenger* " Report on Deep Sea Deposits." [7] These authors state that glauconite " appears to be most abundant about the lower limits of wave, tidal, and current action, or in other words, in the neighborhood of what we have termed the mudline surrounding continental shores." In the next sentence they pass on to describe the occurrence of glauconite " beyond this line, that is to say, in depths of about 200 and 300 fathoms." Mansfield evidently thought these figures referred to the mudline itself, and so quotes them in reporting Murray and Renard's conclusions. But these authors place the mudline, and the zone of greatest glauconite accumulation, at 100 fathoms, or 600 feet (see for example pages 185, 228–229, 253, 321 of the *Challenger* " Report on Deep Sea Deposits "). Even this more moderate depth is consonant with the idea of a far northwestward transgression of the Atlantic coastal plain deposits; and while modern greensands do accumulate to some extent in shallower as well as in deeper waters, and certain very ancient greensands (Cambrian) are believed to have formed in extremely shallow water,[8] the presence of large quantities of Cretaceous greensand near the present eroded inner margin of the coastal plain is strongly suggestive of fairly deep water there. Murray and Renard [9] report that " no typical glauconitic sands have, so far as we know, been recorded in process of formation in the littoral or sublittoral zones."

In concluding this phase of our discussion we may say that the physiographic requirements of the theory of regional superposition are not decisively negatived by stratigraphic and paleontologic evidence. Stephenson [10] while inclined (as we have already seen) to limit narrowly the former northwestward extension of the coastal plain in the New Jersey region, takes no dogmatic stand in the matter; and he frankly recognizes " the difficulty of determining just how far Cretaceous and Tertiary seas may have extended inland across the continent." Darton[11] goes so far as to express his opinion that " the positions of shorelines have varied greatly in Cretaceous and Tertiary times," and speaks of " some fine sediments with remains of life probably indicating considerable depth of water." Clark and Martin[12] place the sedimentation of part of the coastal plain in " seas of moderate depth, probably from 100 to 300 fathoms," a depth much greater than is required by the theory of regional superposition of Appalachian drainage. The testimony of the glauconite seems on the whole favorable to a significant northwestward extension of the formations in which it occurs. But even should one doubt the evidence of beds now available for our inspection, it must be remembered that the theory here set forth does not depend on the testimony of any deposits at present remaining in the far from complete coastal plain record.

Significance of the Fall Zone

As the theory of Appalachian evolution here advanced was taking shape, it became obvious that it must involve a new interpretation of the significance of the " fall line," or better still the " fall zone," bounding on the southeast the crystallines of the older Appalachians. Whereas the current interpretation of this physiographic feature required for the development of the falls one kind of relation between stream profile and upland surface, it was realized that the new theory required a distinctly different relation between these two elements of the topography. At the writer's suggestion a study of this phase of the problem was undertaken by Mr. (now Dr.) George T. Renner, who found that the relations actually existing along typical streams traversing the fall zone were those anticipated on the basis of the new interpretation. As Renner's results have been published the reader is referred to his paper [13] for further discussion of the matter.

History of New England Rivers

We must further be prepared, if we accept the theory of regional superposition of southeast-flowing streams from a coastal plain cover, to make significant changes in most of the history of Appalachian drainage hitherto set forth in the literature by different writers. The case of the Connecticut River is typical of many in New England. As we have already seen, the lower Connecticut is, on the new theory, no longer believed to flow in a deflected course acquired in post-Schooley time, but is interpreted as an ancient remnant of the initial drainage consequent upon the uparching of a blanketed peneplane of pre-Schooley date. The north-south course of the middle Connecticut must be regarded as of more recent origin than the lower course, instead of more ancient as in the view presented by Tarr and favorably regarded by Davis. Similar considerations are involved in interpreting the southeast middle course, and the north-south upper course of the Housatonic. The southwest lower courses of the Housatonic and certain other streams require explanation; and one apparently satisfactory is found if we regard them as recently acquired subsequent courses developed on the crushed zones of faults, or on infaulted strips of weak Triassic or Cretaceous beds later entirely removed. Projected profiles of the country east and west of the lower Housatonic reveal a distinct block-fault topography in the crystalline mass, the apparent fault scarps or fault-line scarps in some cases appearing as extensions of known northeast-southwest faults in the Triassic Lowland of Connecticut. A more intensive study of this phase of the problem is now in progress.

REFERENCES

[1] Lloyd W. Stephenson: Personal communication.

[2] Lloyd W. Stephenson: "Unconformities in Upper Cretaceous Series of Texas," *Bull. Amer. Assoc. Petroleum Geologists*, 13^2, 1323–1334, 1929, see p. 1333.

[3] *Ibid.*, p. 1332.

[4] *Ibid.*, p. 1332.

[5] C. Wythe Cooke and Lloyd W. Stephenson: "The Eocene Age of the Supposed Late Upper Cretaceous Greensand Marls of New Jersey," *Jour. Geol.*, Vol. 36, 139–148, 1928, see p. 148.

[6] George Rogers Mansfield: "Potash in the Greensands of New Jersey," *U. S. Geol. Surv.*, Bull. 727, 1–142, 1922, see p. 138.

[7] John Murray and A. F. Renard: "Report on Deep Sea Deposits," *Voyage of H. M. S. Challenger*, 525 pp., 1891, see p. 383.

[8] W. H. Twenhofel *et al.*: "Treatise on Sedimentation," 661 pp., Baltimore, 1926, see p. 339.

[9] John Murray and A. F. Renard: "Report on Deep Sea Deposits," *Voyage of H. M. S. Challenger*, 525 pp., 1891, see p. 383.

[10] Lloyd W. Stephenson: Personal communication.

[11] N. H. Darton: Personal communication.

[12] William Bullock Clark and George Curtis Martin: "The Eocene Deposits of Maryland," *Md. Geol. Surv.*, Volume on Eocene, 19–92, 1901, see p. 57.

[13] George T. Renner: "The Physiographic Interpretation of the Fall Line," *Geographical Review*, Vol. 17, 278–286, 1927.

CHAPTER FIVE

HISTORY OF PENNSYLVANIA DRAINAGE

General Considerations

Acceptance of the theory of regional superposition would involve radical changes in the drainage history of New Jersey and Pennsylvania as set forth by Davis [1] in his classic essays on these two regions. In particular the complicated succession of events invoked to account for the present stream courses of Pennsylvania would be replaced by a history far more simple. For however complex may have been the ancient drainage evolution on the Appalachian folds, the theory here advanced supposes that all results of such history were extinguished when the sea encroached upon the old pre-Schooley surface and buried it under a mantle of sediments. All the streams considered by Davis to be descendants of original Permian rivers consequent on the constructional topography of that time must under the theory of superposition be regarded as later in age and of different genesis. The present drainage is held to date from the initiation of southeast consequent rivers upon the uparched coastal plain sediments. The superposition of these streams upon the pre-Schooley peneplane, which latter presented alternate bands of resistant and nonresistant rocks effectively beveled and ready for speedy etching by subsequent streams, provided the conditions requisite for development of the present stream pattern.

Manifestly one test of the validity of the theory of superposition must be its competence to explain those peculiarities of Appalachian drainage to account for which other theories have been invoked. If it can explain them more simply and more satisfactorily than does a more complicated succession of events, this may perhaps be counted in its favor. While we cannot claim to have established the validity of the theory here set forth by tests sufficiently rigorous to exclude alternative interpretations, its application to some of the difficult problems of Pennsylvania drainage gives results which seem to justify its retention and further examination as a working hypothesis.

Davis' "Rivers and Valleys of Pennsylvania" will ever rank as one of the most brilliant examples of close deductive reasoning to be found in physiographic literature. Fully to appreciate it one must not merely comprehend the utter vagueness of previous ideas on the erosional history of the Appalachians and on the behavior of rivers in general; he must also gain a clear mental picture of the detailed geological structure of eastern Pennsylvania, and then follow step by step the shifting images of changing forms developed at different levels upon a variety of structures. Who does this will have no doubt that the great value of the work lies in deducing the normal evolution of drainage in a region of folded mountains like the Appalachians. This value remains, whether or not one accepts the histories set forth as the most probable explanations of the present drainage features of Pennsylvania.

Davis himself was fully aware of the complicated and sometimes doubtful character of his explanations of specific Pennsylvania features, and was conservative in drawing conclusions from them. He repeatedly cautions the reader in such terms as:

It seems incredible that the waste of the valley slopes should allow [certain suggested captures to take place]; . . . I am free to allow that this has the appearance of heaping hypothesis on hypothesis. . . . Like several of the other explanations here offered, it is presented more as a possibility to be discussed than as a conclusion to be accepted. . . . It may be difficult for the reader to gain much confidence in the efficacy of the processes suggested. . . . The history of the Susquehanna, the Juniata, or the Schuylkill is too involved with complex changes, if not enshrouded in mystery, to become intelligible to any but advanced students. . . . This is too complicated, even if it should ever be demonstrated to be wholly true, to serve as material for ordinary study.

Not all who have followed in Davis' steps have been as cautious in adopting, as he was in presenting, his interpretations of the more difficult drainage histories.

It is important to note that the theory of superposition here presented was not rejected by Davis as incompetent to explain the features described by him. On the contrary, it does not figure in his essay among the hypotheses receiving examination. True, in a prefatory section he defines superposed streams; and in a later section he appeals to local superposition from a river flood plain to account for the curious position of the Susquehanna where it cuts across the ends of two synclines north of Harrisburg. But his discussion is throughout a contrast between two major hypotheses only: (1) that antecedent streams held their courses northwestward across the growing Appalachian folds, the larger of these streams persisting to the present time although suffering a reversal of direction to give their southeast flow; (2) that a new system of consequent streams originated on the growing folds, with master consequents flowing toward the northwest; these master streams later suffering reversal toward the southeast accompanied by heavy losses to subsequent streams growing rapidly headward along belts of weak rock. No attempt was made to see whether the observable peculiarities of Pennsylvania drainage could be explained by early general superposition of a southeast drainage, followed by long continued readjustments of stream courses to structure during the Schooley and later erosion cycles. Thus the way remains open for consideration of a third working hypothesis.

The Davisian interpretation of Pennsylvania drainage (hypothesis 2 of the preceding paragraph) required that author to reconstruct the topography which the completed Appalachian folds would have if no erosion had accompanied the deformation. This reconstruction necessarily involved three postulates: (1) that there was essential constancy in the thickness of the Paleozoic sediments over the entire area in question; (2) that the dips and folds of the beds now exposed at the surface of the ground may in a general way be projected upwards into the air in order to restore the form of the eroded beds; (3) that by reconstructing from the completed folds the form which the country would have if unworn, we gain a sufficiently definite picture of the form through which it actually passed at the time of initial and progressive folding. Opinions may differ as to how safe are these postulates. My own feeling is that postulates (2) and (3) in particular introduce large possibilities of error into Davis' subsequent deductions. Be that as it may, the theory here suggested dispenses with all three postulates and the long, complicated drainage history based upon them. It substitutes a relatively short and simple history, dating from late Cretaceous or early Tertiary instead of from Permian time. Since Davis recognizes that the Susquehanna below Harrisburg and the New Jersey rivers southeast of the crystalline highlands were superposed from a former coastal plain cover, the suggested theory involves no new processes and no new geologic features. It merely extends farther inland two geologic features well known to have had some extension in that direction, and invokes for the folded Appalachian belt processes already described as having operated in the belt of country immediately southeast. True, it also places the coastal plain cover upon a peneplane earlier than that recognized by Davis, and assigns his "Cretaceous" peneplane (Schooley) to the Tertiary period; but for both these changes there is ample warrant in the field evidence.

It is important that the reader keep clearly in mind, during his perusal of the pages which follow, the essential fact that the theory of superposition from a surface earlier than the

Schooley necessarily implies profound modification of the superposed drainage by the growth of countless subsequent streams along weak rock belts, with resulting diversion of streams less favorably situated. Because the earlier Fall Zone peneplane had, according to our theory, beveled the Appalachian folds, the Schooley cycle began with conditions far more favorable to rapid growth of subsequent streams than is implied in Davis' interpretation of the drainage history of Schooley time. Qualitatively, the drainage adjustments following superposition will be much the same as those described by that author; and I know of no adjustments described by him which cannot rationally be explained by the theory of superposition. Quantitatively, the advantages seem to lie with the latter theory, which in a shorter space of geological time presents superior opportunities for extended development of subsequent rivers.

As is implied in the preceding paragraph, there are among the many peculiarities of Pennsylvania drainage discussed by Davis some which may quite satisfactorily be explained by the theory of subsequent readjustment from ancient consequent stream courses, even though a simpler history might be predicated on the theory of superposition followed by subsequent adjustment in the Schooley and later cycles. In this category we may mention the existence of lateral water gaps near but not at the apices of synclinal ridges and the avoidance of the Broad Top Basin by certain of the headwaters of the Juniata River. But there are four problems or groups of problems in which the superiority of the theory of superposition seems almost decisive. To these we may now direct our attention.

Reversal of Pennsylvania Drainage

Let us first consider the remarkable reversal of drainage necessarily involved in the Davisian interpretation of Pennsylvania rivers. That author first shows that the drainage consequent upon the folded structures joined master consequents draining northwestward (Fig. 13, A). Since the master streams now drain southeastward, there was presented the difficulty of reversing the direction of main stream flow (Fig. 13, B). Two possible explanations, not mutually exclusive, are considered: depression of the area of Newark deposition with resulting reversal of the central segments of the consequent rivers, followed by later reversal of other parts of the streams; and progressive capture of the headwaters of the northwest-flowing streams by others having southeast courses. The difficulties of the capture theory are so great that Davis prefers the first alternative, although he points out that this necessitates further explanation of the fact that the streams, once turned southeast, were not again reversed to their original northwest courses when the Newark deposits were strongly tilted in that direction.

The reversal of major streams after these are once firmly established on a land mass, while that land mass remains continuously exposed above sea level with the streams continuously operating upon it, seems to the present writer a drastic procedure, although he would not like to see it excluded from the theoretical possibilities deserving careful consideration. That Davis shared this feeling is indicated by his frank admission that the theory of reversal is " the least satisfactory of the suggestions " presented in his analysis of Pennsylvania drainage. To some it may seem an equally heroic procedure to wipe out the earlier drainage by a marine transgression, and then derive the southeast drainage by superposition from a southeast-sloping Atlantic coastal plain (Fig. 14). But in favor of this latter theory one may urge: first, that such marine transgression is known to have taken place repeatedly in many parts of the world; second, that there is general agreement that the process operated to some extent in the region in question, and that to its intervention the southeast courses of major streams just outside the folded Appalachians are ascribed by Davis and others; third, that superposition requires no assumptions as to earlier

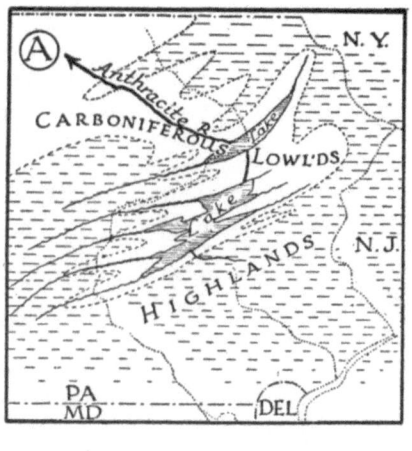
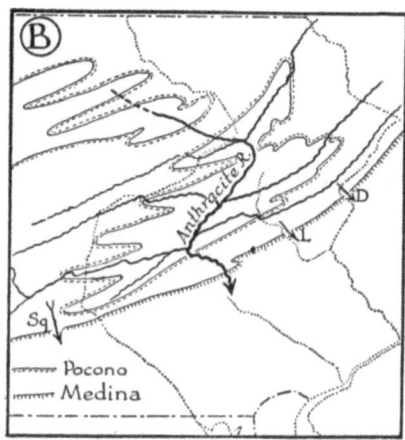
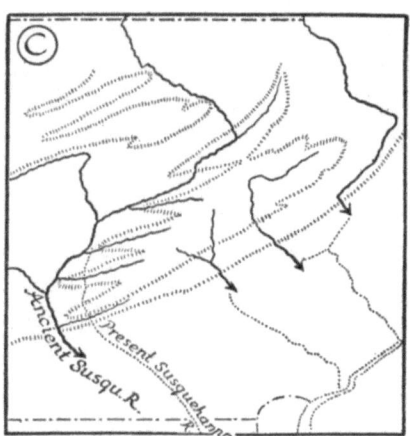
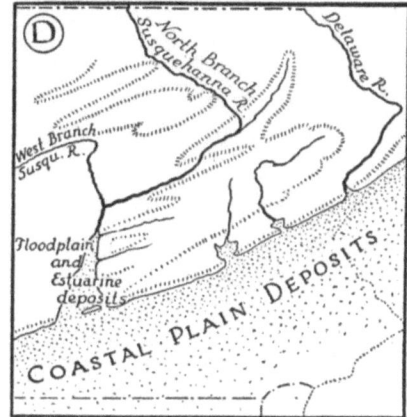

FIGURE 13

Evolution of Eastern Pennsylvania Drainage, According to the Theory of Reversed Consequent Streams

A and B are based on sketches by Davis. E represents the present drainage. C and D are interpolated by the present writer on the basis of Davis' text.

drainage in the region, dates from a comparatively recent period in geologic history, and involves no process of doubtful efficacy.

Disappearance of the Anthracite River

In the case of the major consequent river which Davis supposes to have occupied eastern Pennsylvania, it is not sufficient to reverse it from a northwest to a southeast course. Having created the great Anthracite River (Fig. 13, A), and then reversed it (Fig. 13, B), he must next get rid of it; for today there is, excepting part of the Schuylkill, no large stream where the Anthracite is presumed to have once held sway. Capture by the growing headwaters of southeast-flowing streams more favorably situated is appealed to in order to effect the necessary disappearance. But capture of a great master river by the headwater tributaries of smaller streams demands very special favoring circumstances. Davis gives advantage to the principal actor in the supposed piracy, the embryonic Susquehanna (Sq, Fig. 13, B), by assuming for it a former location on weak rocks west of the river's present course across the two synclinal folds of resistant Pocono sandstone north of Harrisburg. But this assumption leaves the small stream still stranded across the massive Medina (Tuscarora) barrier, a situation little favorable to the vast conquests (Fig. 13, C) the growing Susquehanna is supposed to have effected. While Davis believes the process of capture to be qualitatively correct, he apparently recognizes that one may with some reason doubt its quantitative sufficiency. We are even inclined to doubt the qualitative sufficiency of a theory which ascribes vast conquests to a short, insignificant stream stranded across one of the most formidable hard rock barriers in the whole of the Appalachian province.

If the theory of superposition from a coastal plain cover (Fig. 14, A) be accepted we have no need to concern ourselves with the initiation, reversal, and extinction of a supposed Anthracite River. To be sure, a southeast-flowing superposed stream might be dismembered and diverted to other directions through capture by more favorably situated subsequents (Fig. 14, B and C) during the long Schooley and later erosion cycles, provided always that the conditions required for such captures

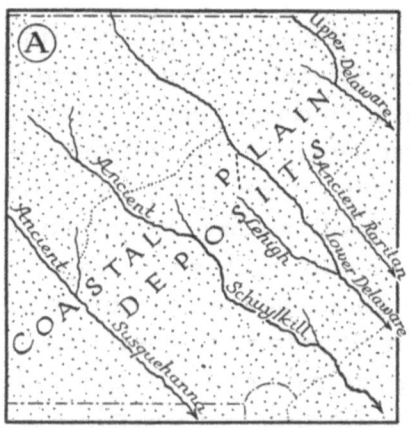

FIGURE 14

Evolution of Eastern Pennsylvania Drainage, According to the Theory of Regional Superposition

A shows initial consequent drainage of the region, and B indicates the action of stream erosion after the streams reached the hard rock barriers. C shows present stream drainage.

existed. But on the theory of superposition the mere absence of a southeast-flowing stream at one place does not of itself present any more of a problem than does the presence of such a stream at another place.

Superposition of the Susquehanna Above Harrisburg

The fact that the master southeast-flowing stream of today, the Susquehanna, is clearly superposed across the ends of two synclinal folds involving the resistant Pocono sandstone north of Harrisburg (Fig. 13, E), presents a puzzling problem to any one who would derive the drainage by subsequent readjustments of an early consequent drainage. Davis expresses this difficulty in the following terms:

There is, however, one apparently venturesome postulate that may have been already noted as such by the reader; unless it can be reasonably accounted for and shown to be a natural result of the long sequence of changes here considered, it will seriously militate against the validity of the whole argument. . . . It was . . . assumed that the embryonic Susquehanna gained possession of the Siluro-Devonian lowland drainage by gnawing out a course to the west of these synclinal points. . . . It is therefore important to justify the assumption as to the more westerly location of the embryonic Susquehanna, and afterwards to explain how it should have since then been transferred to its present course.

Justification of the assumed westerly location of the embryonic Susquehanna is found in the belief that "in the absence of any antecedent stream, it must have lain there." This, however, fails to take account of possible early superposition of the Susquehanna across the folds. To explain the supposed transfer of the river to its present position, Davis assumes that the Susquehanna, grown great by capture of formerly northwestward-flowing drainage, developed a broad flood plain which possibly at this locality merged into estuarine deposits of the encroaching Cretaceous sea (Fig. 13, D). This flood plain-estuarine cover is assumed to have blanketed the hard rock ridges (Pocono) of the synclines, which doubtless were somewhat embossed upon the Schooley peneplane surface, and thus to have permitted the wandering river to become locally superposed across them. In support of this interpretation Davis points to the fact that the lower courses of certain tributaries to the Susquehanna show just such marked deflections as are frequently observed where tributaries join master streams on flood plains of today.

The theory of drainage development advanced in the present volume assumes that the Susquehanna was directly superposed upon the synclines from an ancient coastal plain cover (Fig. 14). The apparent fact of superposition, which on the theory of readjusted consequent drainage presented a difficulty requiring a special explanation, now becomes an essential element of the picture. The shorter and simpler history of regional superposition dispenses with the need for assuming an early westerly location of the Susquehanna, its growth to a great river by an extended series of captures despite the obstacle of the Medina ridge, the development of widespread flood plain-estuarine deposits at a certain locality, and extensive shifting of the river on those deposits. Greater simplicity in a theory is, of course, no guarantee of its validity, since Nature sometimes takes the more complicated pathway to a given end. Nevertheless, a simple explanation which adequately accounts for all observed facts is more apt to be correct than one which involves a long and highly specialized history and explains only part of the facts. Davis' theory left unsolved the question as to how the embryonic Susquehanna acquired its course across the Medina ridge, a question no less vital than that involving the Pocono ridges. The theory of regional superposition reasonably accounts for the course of the Susquehanna across all the ridges.

The apparently deflected tributaries to which Davis called attention present no difficulties for the theory of regional superposition. They may have been incised from flood plain courses acquired on the coastal plain cover or later during the Schooley cycle, without involving estuarine conditions or any shifting of the Susquehanna from a former position farther west. But it is not necessary to regard all those parts of stream courses in question as deflected. They may, in many cases, be remnants of the early drainage which was consequent upon the coastal plain cover and tributary to the master consequent Susquehanna, the upper portions of these tributaries having lost their superposed consequent locations through the normal processes of adjustment to structure during the long Schooley cycle. In this connection it should be noted that in a region of Appalachian structure the lower portions of such superposed tributaries are less apt to be diverted by growing subsequents than are portions farther removed from the master stream.

Two facts brought to light since Davis wrote his classic essay on the " Rivers and Valleys of Pennsylvania " make it especially difficult to accept his interpretation of the Susquehanna history. The discovery that the Schooley peneplane is younger instead of older than the Cretaceous coastal plain renders inadmissible the idea that the Cretaceous sea advanced over the Schooley surface to the vicinity of the synclinal folds upon which the Susquehanna is imagined to have been superposed from a flood plain-estuarine cover of post-Schooley age. It is still permissible, however, to entertain the possibility of local superposition, from either flood plain or estuarine deposits of Cretaceous date, upon a surface older than the Schooley; or to invoke local superposition, from a flood plain unrelated to Cretaceous estuarine deposits, upon the Schooley surface in post-Cretaceous time.

Of more serious import is the recent discovery made by Ver Steeg,[2] while studying the wind and water gap phase of this problem, that the crests of the ridges for long distances descend gently, both from the east and from the west, toward the water gaps by which the Susquehanna traverses the folds. We see no escape from the conclusion that gradual descent of the Schooley surface (or of low monadnock ridges slightly embossed on that surface) toward the Susquehanna indicates long occupancy by the river of its present position. The conditions actually observed are such as should obtain if the Susquehanna held its present course throughout the whole of the Schooley cycle, and thus support the hypothesis of regional superposition which has guided our Appalachian studies. Had the great river formerly pursued a course west of the synclinal apices, and done its part in fashioning the Schooley peneplane while in that position, as Davis inferred, we should expect any prolonged and definite descent of ridge crests to be more or less continuously westward, although local variations in slope might be expected where minor streams traversed the ridges. Later shifting of the river on a flood plain or on estuarine deposits blanketing the Schooley surface, and local superposition in its present course, should give water gaps cutting the ridges in complete discordance with the ancient regional slopes of ridge crests. Instead, we find regional slopes of ridge crests nicely adjusted to the present position of the river, a strong indication that the river held this course while those slopes of the Schooley surface were being fashioned.

If it be agreed that the observations of Ver Steeg indicate occupancy by the Susquehanna of its present position across the synclinal folds during Schooley time, then we must further admit that the supposed growth of the Susquehanna from a small stream to a great river through capture of the hypothetical Anthracite River becomes increasingly improbable. Davis saw the necessity of placing the embryonic Susquehanna on weak rocks west of the resistant synclinal ridges in order to make possible the supposed capture, since a small stream hampered by the task of carving a pathway through several great hard rock barriers could scarcely make vast conquests from the mighty Anthracite. Evidence that the Susquehanna has long held its present course thus militates strongly against the theory of an embryonic stream growing great by piracy, although it does not exclude the possibility of significant captures by subsequent branches of a great Susquehanna River, early superposed from a former coastal plain cover.

Davis' Doubtful Cases

Under the caption "doubtful cases" Davis writes:

It is hardly necessary to state that there are many facts for which no satisfactory explanation is found under the theory of adjustments that we have been considering.

Then follows a brief catalogue of some of the features of Pennsylvania drainage which he found it difficult to reconcile with his theory of stream adjustments from earlier consequent courses. Among these he mentions first the location of the Susquehanna across the ends of the Pocono synclines north of Harrisburg, on the ground that some might hesitate to accept his suggested explanation of local superposition from a flood plain-estuarine cover. We have already seen that the theory of regional superposition more easily accounts for the observed facts.

The second category of "doubtful cases" included series of water gaps definitely aligned, usually in a northwest-southeast direction. For some of these Davis could suggest a reasonable origin on the basis of his theory; but for others he records that no satisfactory explanation could be found. We have considered this phenomenon in an earlier paragraph and have seen that such alignment of gaps is not merely a normal accompaniment of the theory of regional superposition — it is an inevitable consequence of that theory.

As a third item in his list Davis notes that "the location of the upper North Branch of the Susquehanna is also unrelated to processes of adjustment as far as I can see them." Apparently reference is here made to the relatively straight northwest-southeast course of the stream from the New York border to the Wyoming syncline southwest of Scranton (Fig. 13, E). It will be noted that this part of the river, shown by dotted lines in diagram B of Figure 13, lies far to the northeast of the supposed course of the reversed "Anthracite River" represented by a heavy black line in the same figure. In diagram C the upper Anthracite, now become the upper North Branch of the Susquehanna, is arbitrarily shifted northeast to the present location of the stream. This change is made not because Davis gives any map or diagram showing such transfer of drainage (diagrams C and D of Figure 13 are interpolated by the present writer in the absence of drawings illustrating the appropriate parts of Davis' text), but because such change must have occurred in some manner under the Davisian interpretation. It seems to be this transfer of drainage which Davis could not adequately account for on the theory of normal adjustment of streams consequent on the folded structures. On the theory of regional superposition of southeast-flowing streams consequent on a coastal plain (Fig. 14, A), which streams later suffered partial adjustment to the structures on which they were superposed (Fig. 14, B and C), the remarkably long, relatively straight course of the upper North Branch of the Susquehanna presents no problem. It is not only a normal, but also a necessary, part of the drainage history.

Fourth among the "doubtful cases" Davis cites the following:

The great area of plateau drainage that is now possessed by the West Branch [of the Susquehanna] is certainly difficult to understand as the result of conquest.

The plateau drainage in question is composed chiefly of southeast-flowing streams. These are tributary to larger streams flowing northeast or southwest parallel to the structural belts of the folded Appalachians, and apparently developed on weak rock layers either now or formerly marking their courses. Both portions of the drainage find ready explanation under the theory of regional superposition: the southeast-flowing streams as superposed consequents let down upon the underlying structures from the former coastal plain cover; the northeast- and southwest-flowing streams as subsequents developed in the normal course of adjustments during the Schooley and later cycles.

The fifth and last of his doubtful cases is thus presented by Davis:

The two independent gaps in Tussey's Mountain, maintained by the Juniata and its Frankstown Branch below Tyrone,* are curious, especially in view of the apparent diversion of the Branch to the main stream on the upper side of Warrior Ridge (Oriskany), just east of Tussey's Mountain.

The pattern of the Frankstown Branch just above its junction with the Juniata northwest of Warrior Ridge does suggest the possibility of capture. But an equally plausible interpretation of the pattern is found in the development of a single loop or meander of the Branch, by which the lower few miles of its course shifted southward from a former path due eastward to an original junction with the Juniata at or near the same point as the present junction. The topography is favorable to the latter interpretation, and gives no clear indication of a former independent course of the Branch across Warrior Ridge, followed by later diversion to the present junction. Theoretically the capture seems improbable: first, because it would require that an insignificant brook only two or three miles long should reduce its headwaters below the level of the lower course of a relatively large river which apparently was no more impeded by hard rock barriers than was the master stream to which the brook was tributary; and second, because the supposed capture involves diversion of a large river, the Frankstown Branch, into a neighboring river similarly situated, the Juniata, when the latter apparently enjoyed little if any advantage as to volume of water, which under the circumstances would seem to be the only factor capable of inducing capture.

If we assume superposition of southeast-flowing streams from an ancient coastal plain cover, the capture suggested by Davis

* See the Tyrone (Pennsylvania) topographic quadrangle for the detailed features cited in this discussion.

would be as feasible as, but no more feasible than, it would be under the theory of normal adjustments of streams consequent on the folded structures. A more plausible interpretation would be that two streams flowing southeast on the coastal plain cover formed a junction northwest of the present site of Warrior Ridge, but southeast of Tussey's Mountain. Superposition on the buried Appalachian structures would give separate but closely spaced gaps in Tussey's Mountain just above the point of junction; but a single gap where the combined waters flowed as a single stream through Warrior Ridge. If it be granted that such a history is reasonable (and it seems to me certain that such cases must occur under the theory of regional superposition), the persistence of the two closely adjacent gaps to the present time is not a phenomenon difficult to explain. The large Juniata would conquer the territory on its left, the large Frankstown Branch the territory on its right. In the narrow strip of territory between the two, where the rivers were closely adjacent, only insignificant streams could develop. These would be powerless to undermine and tap either of the large rivers, so long as neither the Juniata nor the Frankstown Branch acquired an enormous advantage over its rival. Such advantage seems to be lacking, and in its absence the persistence of both gaps through Tussey's Mountain must be regarded as inevitable.

It thus appears that the theory of regional superposition of streams from a coastal plain cover resting on a peneplane of pre-Schooley (pre-Kittatinny) date does more than account in a reasonable and simple manner for the features of Pennsylvania drainage derived by Davis in a more complicated manner: it offers a rational explanation for all five categories of facts which under the Davisian theory remained unexplained. We seem justified in saying that the theory of regional superposition, subjected to the rigid test of deducing its necessary consequences as applied to the interpretation of complex drainage problems in Pennsylvania, emerges from the trial stronger rather than weaker, and is therefore entitled to an increased measure of confidence as a fruitful working hypothesis.

Relation of Streams to the Nittany Arch

Before leaving the subject of Pennsylvania drainage it is desirable to consider for a moment drainage relations in Center County and adjacent areas. This is the region of the great Nittany anticlinal arch (marked N on Davis' sketch maps of the region) lying northwest of the Broad Top synclinal basin (marked BT on Davis' maps). In comparing the conditions of this region in Permian time with those of the present, Davis [8] says:

The Nittany district, then a highland, is still a well-marked divide, although now a lowland.

If the present existence of a divide on the unroofed Nittany arch were proof of the persistence of such a divide since Permian time, we would have in this fact an indication that the former coastal plain cover did not extend as far inland as the Nittany region.

It is not difficult, however, to account for a divide in this region on the assumption that the Fall Zone peneplane beveled the arch at a higher level than the present, and that a Cretaceous coastal plain cover completely extinguished the former topography, including the divides of earlier times. We do not know at what altitude above the present ridge crests the Fall Zone peneplane would have cut the folded structures. But it is permissible to suppose that there would have been ample opportunity during the long Schooley cycle for resequent streams to develop a new divide along the arch on any one of such resistant beds as the Pottsville, Pocono, or Medina.

The nature of the divide suggests some such history. It is far from being a perfect divide. The Juniata cuts clear across the arch not far south of its central portion, taking a remarkably direct southeast course which is prolonged straight across the northern end of the Broad Top syncline and on across the next anticline and the next syncline to the southeast. Only rarely, as just before entering the second anticline, is there deflection of the Juniata parallel to the structure for two or three miles. For thirty-five miles this stream pursues a southeast course transverse to the folds, a phenomenon which does not seem quite satisfactorily accounted for by the succession of drainage adjustments inferred by Davis when discussing avoidance of the Broad Top Basin by the Juniata headwaters.[4] Farther northeast Spring Creek cuts obliquely across the arch from southeast to northwest. Beyond this point, to the eastward, the arch divides into two minor anticlines with other subordinate wrinkles. Both of these minor folds and some of the wrinkles are repeatedly traversed by small streams which cut clear across them from southeast to northwest, or from northwest to southeast. Still farther east both minor folds of the great arch, and the next anticline to the south, are trenched by the West Branch of the Susquehanna following a course slightly east of south: a course remarkably direct across the structures except for a loop suggesting minor shifting or partial adjustment where the eastern end of the northern fold plunges downward east of Williamsport.

If we imagine the original folds restored, it will appear that while the Nittany arch is in places a divide, the most striking characteristic of the drainage is its prevailing indifference to the constructional topography of earlier times. The observed relations are such as should occur as the result of regional superposition: the larger streams traverse the structures in a general southeast direction; some small streams have similar directions, perhaps as the result of persistence of superposed southeast courses, more likely as the result of later adjustments during the long Schooley cycle; other small streams have northwest courses, probably for the reason last mentioned; while small streams of both groups traverse the axes of the folds in the manner already described. The prominence of southeast drainage in the area northwest of the Nittany arch is further indication of regional superposition of streams from a coastal plain cover which extended clear across the arch and over a portion at least of the Appalachian plateau beyond.

REFERENCES

[1] William Morris Davis: "Rivers and Valleys of Pennsylvania," *Nat. Geog. Mag.*, Vol. 1, 183–253, 1889. *Geographical Essays*, 413–484, Boston, 1909. "The Rivers of Northern New Jersey with Notes on the Classification of Rivers in General," *Nat. Geog. Mag.*, Vol. 2, 81–110, 1890. *Geographical Essays*, 485–513, Boston, 1909.

[2] Karl Ver Steeg: "Wind Gaps and Water Gaps of the Northern Appalachians, Their Characteristics and Significance," *N. Y. Acad. Sci., Annals*, 32, 87–220, 1930, see pp. 135, 182–191.

[3] William Morris Davis: "Rivers and Valleys of Pennsylvania," *Nat. Geog. Mag.*, Vol. 1, 183–253, 1889, see p. 225.

[4] *Ibid.*, see pp. 227–229.

CHAPTER SIX

THE RIVERS OF NORTHERN NEW JERSEY

For the student of Appalachian geomorphology Davis' analysis of " The Rivers of Northern New Jersey "[1] is second in importance only to his classic essay on the " Rivers and Valleys of Pennsylvania." In his discussion of New Jersey drainage this author introduces certain evidence and arguments which have been widely accepted and which are not in harmony with the theory of regional superposition of Appalachian rivers set forth in earlier parts of the present volume. One cannot, therefore, adopt this latter theory unless he is also prepared to accept, as one of its necessary consequences, some revision of the current interpretation of northern New Jersey geomorphology. It is therefore necessary to turn our attention for a time to the New Jersey region.

The crystalline "highlands" of New Jersey (Fig. 15) are divided into long narrow strips of upland by a series of equally long and somewhat narrower valleys. These valleys trend northeast-southwest, and are eroded on infaulted or infolded belts of weak limestone and shale. A typical example is the Musconetcong Valley (Fig. 16) and Davis stresses the perfection of adjustment to weak rock structures exhibited by it and its fellows. Such adjustment is reasonably attributed to long opportunity for the streams to seek out and follow lines of least resistance. It is inferred that during much of Jura and Cretaceous time this process was in operation, and that the effect was never destroyed by later overlap of the Cretaceous coastal plain across the highland area. In Davis' opinion " the pebbles,

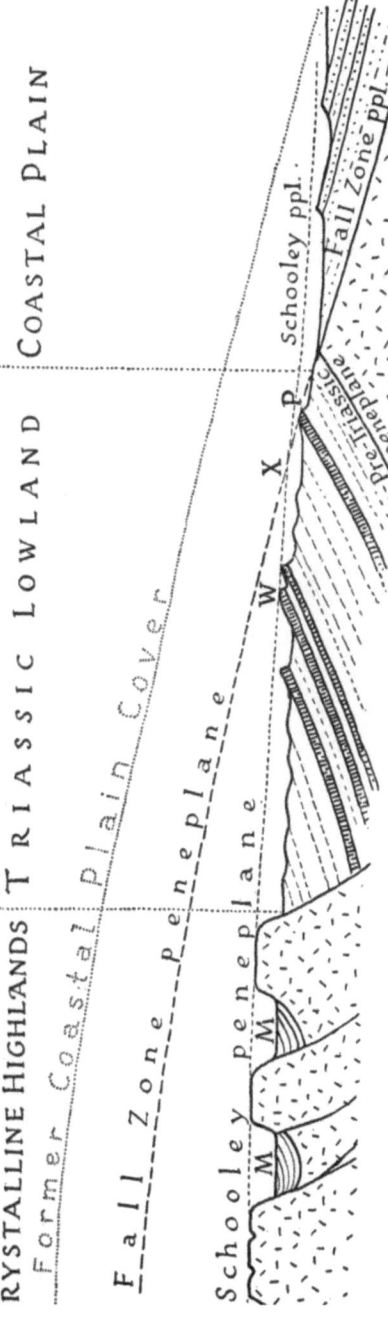

FIGURE 15

Section of Northern New Jersey Showing the Geomorphic Evolution, According to the Theory of Regional Superposition in Pre-Schooley Time

M, M = Musconetcong and similar adjusted streams of the highlands area
W = Watchung Mountains trap ridges
P = Palisades trap ridge
X = Intersection of Schooley and Fall Zone peneplanes

sands, and marls of the Cretaceous series point clearly to the highlands as their source," whereas these same Cretaceous beds " contain practically no Triassic fragments." [2] He therefore believed that while the Cretaceous sea spread westward over the low Schooley peneplane to the eastern margin of the crystalline belt, it did not extend far beyond that margin. Thus the peneplaned Triassic rocks were blanketed with a coastal plain cover (Fig. 18, A), while the peneplaned crystallines (later to become the " highlands ") with their limestone valleys were not so covered. The inevitable result of such a history would be to leave undisturbed the stream adjustments acquired through long ages in the continuously exposed highlands district, but to blot out such adjustments in the recently blanketed Triassic region.

If we now turn our attention to the Triassic belt, especially to that portion where the curved edges of two great warped lava flows project above the surrounding lowland to form the crescentic ridges of First and Second Watchung Mountains (Fig. 15, W), we find conditions quite unlike those of the highlands area. As Davis[3] tells us:

The noteworthy feature of this [Watchung] district is that the small streams in the southern part of the crescent rise on the back slope of the inner mountain and cut gaps in both mountains in order to reach the outer part of the central plain [see Fig. 16]. If these streams were descended directly or by revival from ancestors antecedent to or consequent upon the monoclinal tilting of the Triassic formation, they could not possibly, in the long time and deep denudation that the region has endured, have maintained, down to the present time, courses so little adjusted to the structure of their basins. In so long a time as has elapsed since the tilting of the Triassic formation, the divides would have taken their places on the crest of the trap ridges and not behind the crest on the back slope. . . . Their courses must have been taken *not long ago*,* else they must surely have lost their heads back of the Second Mountain. . . . The only method

---

* Italics are in the original text by Davis.

now known by which these several doubly transverse streams could have been established in the not too distant past is by superimposition from the Cretaceous cover that was laid upon the old Schooley peneplain.

Can the features noted by Davis in northern New Jersey be harmonized with the conception of regional superposition of streams from a coastal plain extending far northwestward beyond the crystalline area of the highlands and resting on a peneplane older than the Schooley? For one seeking an accord between divergent views the source of the coastal plain sediments will present no special difficulty. I take it that the pebbles and sands * mentioned in one of the extracts quoted above are the quartz pebbles and sands ‘ so characteristic of the New Jersey coastal plain series, and that it was not intended to imply that pebbles of crystalline rock or other material had been specifically identified as coming from known sources in the highland formations. If this understanding be correct, then we may find an adequate source for the " pebbles, sands, and marls of the Cretaceous series " in the quartz conglomerates, sandstones, and limestones which figure so prominently in the Paleozoic formations extending far northwest of the highlands. The perfection of adjustment to weak rock belts attained by the Musconetcong and its fellow streams (Fig. 16) in the highlands area is in harmony with the degree of adjustment which we have described for areas farther west in Pennsylvania, and should be expected in New Jersey if the streams there were similarly superposed from a pre-Schooley surface, and so had much of Schooley time, as well as all of post-Schooley time, in which to develop valleys along weak rock belts previously beveled by the Fall Zone peneplane.

* Elsewhere Davis refers to them specifically as "sands and white quartz pebbles."

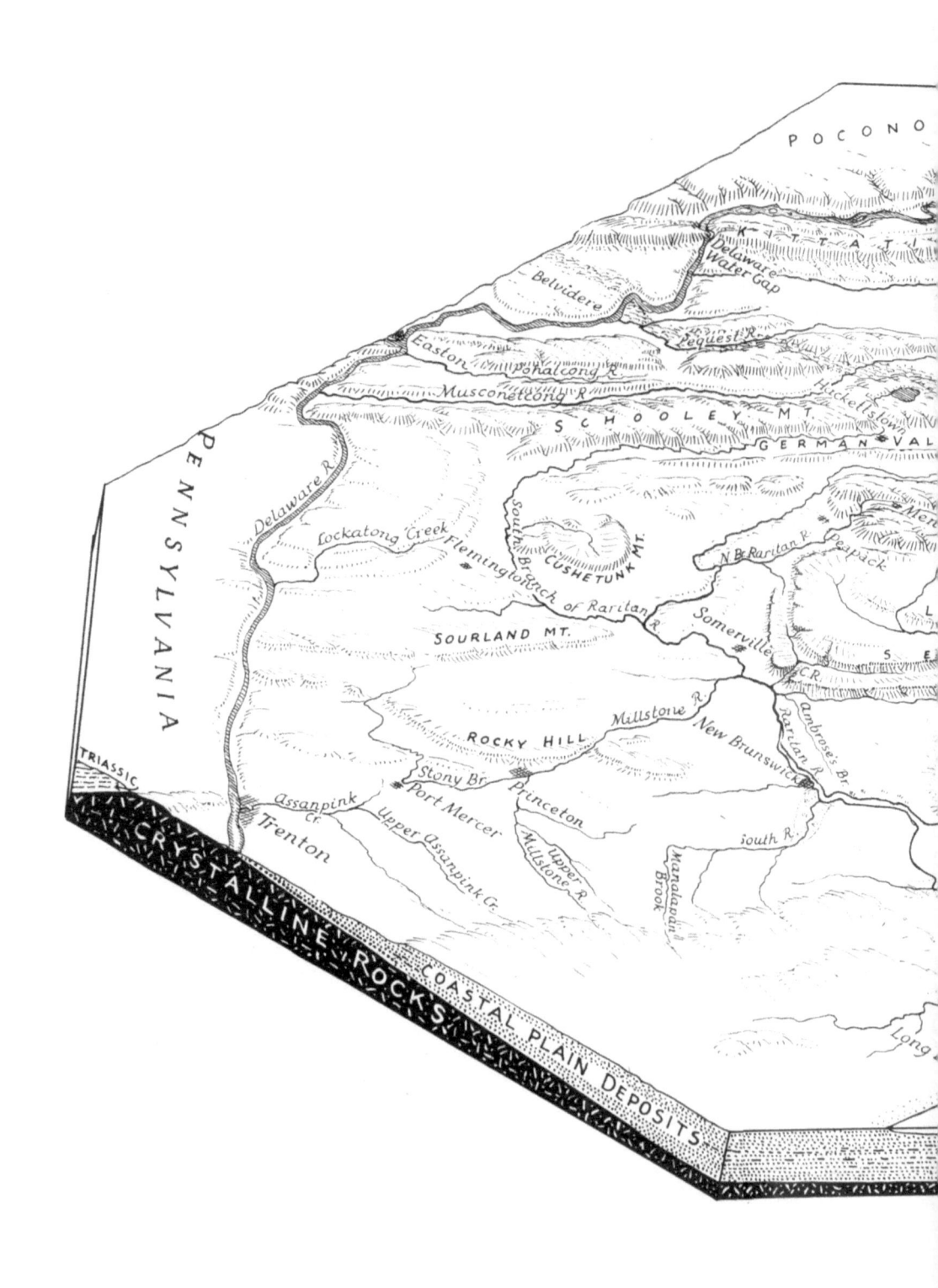

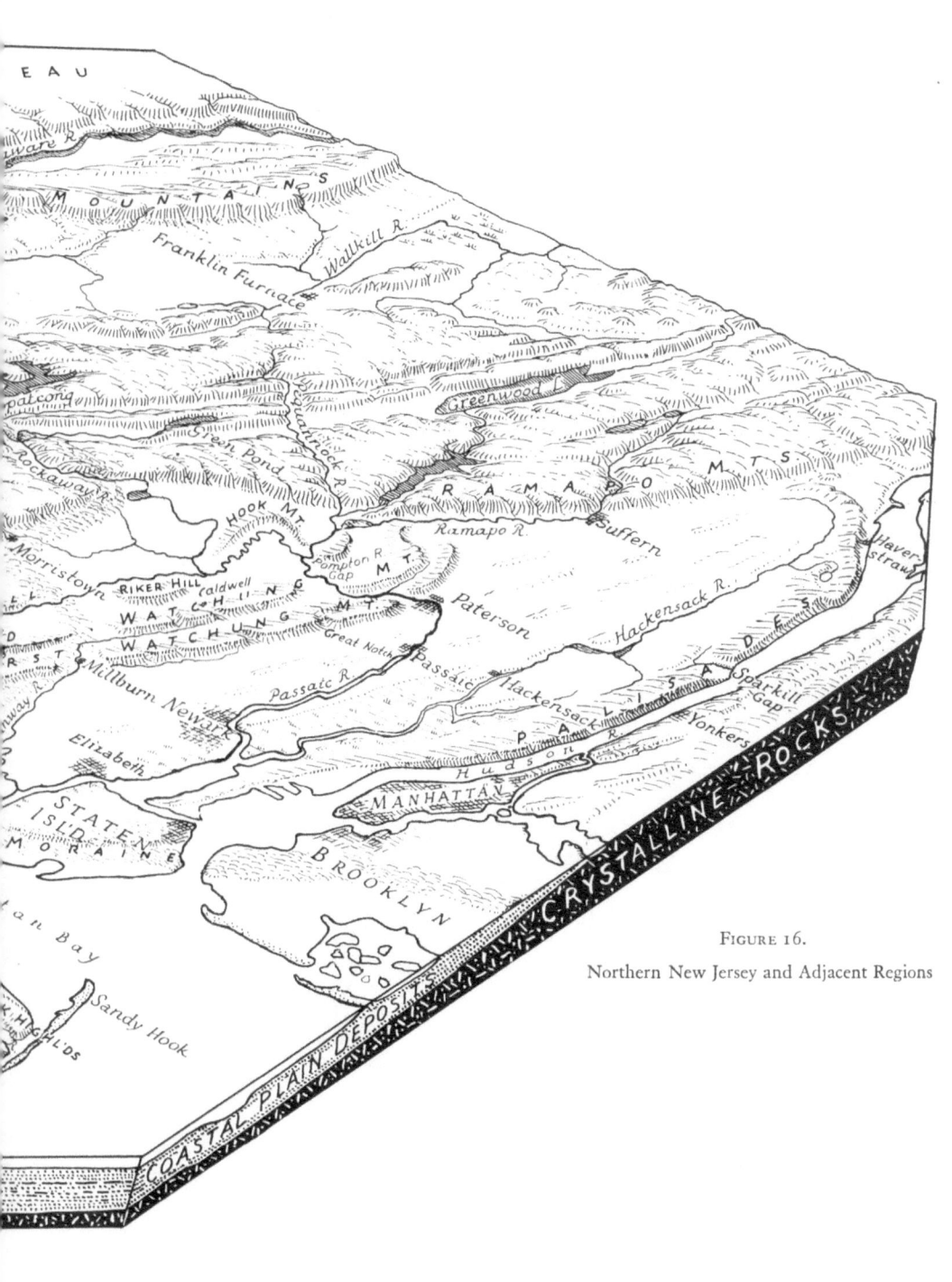

FIGURE 16.
Northern New Jersey and Adjacent Regions

Lack of Stream Adjustment Along the Watchung Ridges

There remains to be explained the imperfect adjustment of streams to structure in the region of the Watchung trap ridges. In the discussion which follows we shall assume that the lack of adjustment is as pronounced as Davis supposed, although it is pertinent to point out that all of the streams in question head on the trap ridge, and not on weak rock. The divides are now located on the resistant barrier, and must have been so located for a long time in the past. That the divides have not been pushed eastward to the main crest of the trap ridge, but still remain on its back slope, may well be due to the fact that headwater ravines of the western streams could not overcome certain advantages enjoyed by the eastern streams. The latter usually have headwater subsequents nicely adjusted to a zone of weakness which causes the double crest of Second Watchung Mountain. Lewis believes that this ridge is made up of two distinct lava flows, separated in most places by a thin bed of Triassic shales. The shale bed constitutes an intermediate zone of weakness which has guided erosion and produced the double crest of the ridge.[5] Not only is there marked adjustment of these streams to structure, but their deep incision in the weak belt, or in the trap beneath it, renders them reasonably immune from capture. Hence Davis' argument loses much of its force. But since it may be held that more perfect adjustment should have taken place before the subsequents were intrenched in the trap, thus shifting the divide to the higher or eastern crest of the ridge, we shall base our discussion on the assumption that Davis was essentially correct in stressing the unadjusted character of the beheaded transverse streams.

The reader must keep in mind the fact that while Davis invoked, as does the present writer, superposition from a coastal plain in part at least of Cretaceous age, Davis believed such superposition occurred " not long ago," " in the not too distant past," and explained the imperfect adjustment of streams in the Watchung region on the theory that there had not been time for such adjustment to take place. It must likewise be remembered that in New Jersey, as in Pennsylvania, Davis placed the coastal plain cover directly on the Schooley peneplane, a feature itself sufficiently recent to constitute a prominent element in the present upland landscape. In the forty years which have elapsed since Davis marked a new epoch in drainage studies by the two essays we have been considering, geologists have pushed the Cretaceous epoch far back in geologic history. Today we hesitate to believe that any topographic form of Cretaceous age could, if continuously exposed to the elements, escape complete destruction during the immensely long periods of denudation which have since elapsed. We recognize that the Schooley peneplane is of more recent date than Cretaceous, and that the Cretaceous coastal plain must have rested on an erosion surface of far greater antiquity than is the Schooley.

Thus while the Davisian interpretation of New Jersey drainage history accorded perfectly with the facts known to geologists at the time of his writings on the subject, and offered a logical explanation of the apparent contrast in degree of adjustment which he noted in streams of the highlands area as compared with those of the Watchung ridges, we may not ourselves appeal to the same interpretation. If our reading of the geomorphic record be correct, the Cretaceous coastal plain covered Triassic and crystalline areas alike, while the Schooley peneplane beveled indifferently the crystallines, the Triassic formations, and the Cretaceous coastal plain deposits, as shown in Figure 15. It remains to inquire, therefore, whether any difference in degree of stream adjustment should be expected on the basis of this history as between streams of the highland area and streams of the Watchung ridges.

As will readily be apparent from Figure 15, the line of intersection (X) of the later Schooley peneplane with the more ancient Fall Zone peneplane is a line of critical importance.

Northwest of this line the Triassic rocks, as well as the crystallines farther northwest, have so far as we know been exposed to denudation throughout much of the Schooley cycle and during all post-Schooley time. Southeast of this line, on the contrary, the Triassic beds were protected by overlying coastal plain deposits all through the Schooley cycle and for a part of post-Schooley time, so that they have been subjected to erosion for a comparatively short period only. If, then, the Watchung Mountains occupied the position P in Figure 15, the imperfect drainage adjustments noted by Davis could readily be accounted for on the general principle invoked by that author, simply by modifying the application of the principle to meet the needs of the geologic history represented in the diagram. It is believed that a portion of the Palisades trap ridge does occupy the position P, southeast of the critical line X; but if the Watchung Mountains (W) are situated to the northwest of this line, then in common with the crystalline area farther northwest they have been subjected to erosion throughout much of Schooley and all of post-Schooley time.

The location of the line of intersection of the Schooley and Fall Zone peneplanes may be fixed with some precision both to the northeast and southwest of the area here under discussion, because in those regions both surfaces are in some measure preserved on resistant crystalline rocks. But in the Triassic region of northern New Jersey the two planes happened usually to intersect on weak rocks, with the result that both surfaces have since been destroyed for some distance on either side of the line of intersection. Under such circumstances the location of this critical line can be determined only approximately. It is believed that Figure 15 correctly represents it as lying some distance southeast of the Watchung ridges, and on this basis we shall proceed to inquire whether there is still a reasonable explanation for the lack of adjustment observed in the Watchung streams.

Obviously we might account for the phenomena in question

by assuming a late Tertiary encroachment of the sea westward across the Schooley peneplane to a line located somewhere west of the Watchung ridges, with concurrent deposition of a sufficient thickness of Tertiary beds to blanket the outcrops of trap and thus later to cause superposition of streams. This would again be the Davisian interpretation, modified to bring it into harmony with the supposed Tertiary age of the Schooley peneplane. This possibility deserves special consideration in view of Salisbury's belief that Miocene (Beacon Hill) beds extended farther inland than the Cretaceous. Salisbury[6] writes:

So far as the arguments which Davis advances are concerned, the younger formation which covered the edge of the old peneplane might as well have been Miocene as Cretaceous. When Davis wrote, the former northern [northwestern] extension of any formation younger than the Cretaceous was not known.

Apparently Salisbury[7] based his conception of a far inland extension of Miocene beds on the occurrence of sands and gravels on remnants of the upland peneplane as far northwest as the Watchung region. There is, however, some doubt as to the marine origin of these beds. In response to a query on this point Dr. Henry B. Kümmel,[8] State Geologist of New Jersey, writes as follows:

I find myself forced to disagree with Mr. Salisbury in his conclusion that the occurrence of these high-level scattered gravels at points considerably northwest of the margin of the Cretaceous is evidence of subsidence and marine incursion late in Tertiary time. It seems to me that they could better be explained as remnants of stream work on a peneplain, which is now largely destroyed.

So far as the writer is aware there is no satisfactory evidence that such a Tertiary coastal plain ever did overlap the Watchung ridges; but it is none the less important to recognize the fact that on this hypothesis we could easily account for the imperfect adjustment of Watchung drainage while at the same time admitting general superposition of Appalachian streams from a Cretaceous coastal plain resting upon an erosion surface of pre-Schooley date.

But it does not seem to me necessary to go so far afield in order to account for the phenomena in question. In the first place, it appears highly probable that the rock types of the Watchung region are less favorable to ready adjustment of streams to structure than are those of the highlands area. The solubility of the limestone belts in the latter district presumably gave subsequent streams there an enormous advantage as compared with subsequents working on the insoluble and somewhat resistant sandstones and shales associated with the Watchung trap sheets. This should mean that in a given period of time adjustment of streams would be less perfect in the Watchungs. In the second place, the inclosed nature of the Watchung basin (Fig. 16) and the possibility that its hard rock rims were nowhere trenched by any stream comparable in magnitude to the Delaware or the Susquehanna, may have operated to retard adjustment within the borders of the basin. A third conclusion relates only to lack of adjustment along First Mountain. The failure of subsequent streams to develop more effectively between First and Second Watchung Mountains (Fig. 16), and thus to collect the drainage of that belt into one or two master subsequents which would long ago have beheaded most of the streams now crossing through the hard rock barrier of First Mountain, must in considerable part be attributed to the narrowness of the weak rock belt between the two trap ridges. Small streams narrowly confined between ridges which from both sides shed abundant coarse debris into the stream channels cannot deepen their valleys as rapidly as do streams located on broader belts of weak rock, and hence cannot prosecute as effectively the work of conquering new territory upon which the process of stream adjustment depends.

Of far greater importance than the three mentioned above is a fourth factor vitally affecting the degree of drainage adjustment in the whole Watchung district. The angle of slope of the Fall Zone peneplane wherever observed in or near the coastal plain province is such that its projection inland, even with an appreciably decreasing slope, will carry it well above the upland remnants of the later developed Schooley peneplane. As shown by Figure 15, this means that in the region of the crystalline highlands a great vertical distance presumably separates the restored positions of the two peneplanes; whereas in the Watchung district (W), only a few miles west of the intersection (X) of the two surfaces, the vertical distance separating the two must be small. In other words, during the Schooley cycle streams of the highlands area enjoyed the advantage of a great vertical distance in which to develop those marked differences of level which make it possible for one stream to capture another and thus to bring about adjustment of stream courses to rock structures. On the other hand, streams of the Watchung district must effect their captures within a short vertical distance or not at all. Analysis of the process of stream capture leaves no doubt that the element of opportunity, expressed in terms of the vertical distance through which competing streams may intrench themselves during their struggle for territorial advantage, is one of the dominating factors in the process of stream adjustment. In this respect the competing streams of the highlands area seem to have enjoyed throughout the Schooley cycle a notable advantage over similarly competing streams of the Watchung region. Nor does this express the whole measure of the highlands streams' advantage. As will appear evident from inspection of Figure 15, this advantage continued during post-Schooley time, since the vertical distance separating the Schooley peneplane from present baselevel is appreciably greater in the highlands area than in the vicinity of the Watchungs.

It thus appears that whether we imagine the Watchung region to lie southeast or northwest of the line along which the Schooley peneplane intersects the earlier Fall Zone peneplane, we find adequate explanation for the lack of perfect stream adjustment in this area. The explanations offered not only seem reasonable and logical as respects their quality, but they appear quantitatively sufficient to account for the extent of differences in adjustment actually observed as between Watchung streams and those of the highlands. Furthermore, these explanations harmonize with that theory of Appalachian evolution which seems to account most simply and most completely for the major drainage features of the Appalachian province as a whole; namely, the theory which derives the southeast drainage by superposition from a coastal plain extending northwest far beyond the crystalline highlands and resting upon a peneplane developed in a cycle antedating the Schooley.

Unadjusted Streams of the Highlands Area

Indeed, as in the case of Pennsylvania drainage, the theory of regional superposition possesses certain advantages in explaining peculiarities of New Jersey drainage. Davis appealed to limited overlap of the coastal plain in order to explain the apparently superposed courses of such minor streams as the North Branch of the Raritan between Mendham and Peapack (Fig. 16), and the Lockatong, a small tributary of the Delaware; but he appealed to faulting [9] as a possible explanation for some at least of the apparently superposed and unadjusted courses of much larger streams like the Rockaway and Pequannock Rivers where these flow southeastward across the structural belts of the highlands (Fig. 16). The original lower courses of these two streams, believed by Davis to be represented by the Rahway and Passaic respectively, were thought by him to traverse the Watchung ridges as a consequence of superposition from a coastal plain cover (Fig. 18); but since the coastal plain was for the most part limited in its extension to the Triassic area, a wholly different explanation had to be invoked for the transverse upper portions of the Rockaway-Rahway and Pequannock-Passaic in the highlands area. The theory of regional superposition accounts at once, and with equal simplicity, for both large and small southeast-flowing streams, for both upper and lower courses of streams which are in both parts transverse to the structure of the region, and for the transverse course of the still greater Delaware across three physiographic provinces in succession.

OFFSET GAPS OF THE SOUTHERN WATCHUNG CRESCENT

One peculiarity of New Jersey drainage noted by Davis is the absence of paired water gaps where minor streams cross the two trap ridges in the southern part of the crescent. Were these streams recently superposed upon the ridges from a southeast-sloping coastal plain which buried them to " a slight depth " only, and were these streams little disturbed by subsequent adjustment, as required by the Davisian interpretation, we should expect to find a clean-cut disposition of water gaps in pairs situated on northwest-southeast lines. Instead, we find the gaps notably offset, with the streams following rectangular or zigzag courses before they can escape through the outer ridge (Fig. 16). Davis explained this peculiarity on the assumption that originally the gaps were in pairs on streams traversing the ridges at oblique angles; " for, as Gilbert has shown, oblique courses across tilted beds, alternately hard and soft, will gradually shift until they follow rectangular courses along the strike of the soft beds and square across the strike of the hard beds." [10]

It will readily be granted that the quality of this explanation is above criticism; and that if we admit the assumption of initially oblique stream courses across the ridges, the quantity of adjustment necessary to transform such courses into the rectangular pattern is not unduly great. It may further be granted that where the crescentic ridges curve round toward the west, the normal northwest-southeast course to be expected in a superposed stream would be markedly oblique to the structural trends. But experiments in restoring the initial drainage of this region, made in coöperation with members of my seminar in physiography at Columbia University, have convinced me that this explanation cannot be applied to all of the offset gaps, unless we assume a degree of stream obliquity out of harmony with the expectable pattern of ordinary superposed consequent streams.

A fuller explanation of the observed facts according to the theory of regional superposition may be offered. According to this theory the major streams of the region and many of the minor streams flowing seaward on the ancient coastal plain were superposed on underlying structures without serious modification of their consequent or resequent southeast courses. But progressive dissection of the coastal plain cover in Schooley time presumably developed a cuesta or cuestas (Fig. 17, A), from the inward facing slopes of which obsequent streams (O, O) flowed northwest into broad subsequent lowlands, which presumably were here not far above the baselevel of Schooley time. It is to be expected that with progressive retreat of cuesta escarpments the subsequent streams will tend to migrate down the exposed surfaces of harder rock, and this implies a corresponding seaward migration of the zone of obsequent streams occupying the inface of each cuesta. It must be recognized, however, that both subsequent and obsequent streams may, under favorable conditions, be superposed upon underlying hard rock structures; and that these superposed courses may long be preserved if local conditions are favorable. Occasional northwest-flowing streams might even persist to the present day, intrenched in the basement rocks and holding courses opposed to the slope of both peneplane and coastal plain; but such examples should be rare. Many more streams may have persisted in such " abnormal " courses for a limited period of time, until they were diverted into more favorable channels by the process of stream adjustment.

If, now, we examine Figure 17 it will be noted that the second or western Watchung trap sheet will be trenched by superposed drainage, before the beveled edge of the first or eastern sheet has been uncovered. Northwest-flowing obsequent streams may cut across the second sheet for a time; a superposed subsequent might even cross and recross it at very oblique angles; or resequents might early gain southeast courses across its exposed edge. Later the same history may be repeated along the more recently exposed edge of the first or eastern sheet; but between the two series of superpositions a significant time interval has elapsed. The interval will be comparatively long if the region was then, as seems probable, not greatly elevated above the position of the Schooley baselevel. But despite the long lapse of time the depth of intrenchment permitted to the streams cannot be regarded as sufficiently great to favor extensive drainage readjustments through capture. On the other hand, the time interval is quite long enough to assure, and on the postulated assumption must assure, appreciable retreat of the cuesta scarp and appreciable reduction of the lowland floor, before the margin of the eastern trap sheet is encountered by the intrenching streams. This means that whether or not any captures have taken place in the interval, streams will have had opportunity to shift their courses so far that parts of streams trenching the eastern sheet may be far out of alignment with those parts which earlier trenched the western sheet. Indeed, different streams, and even different types of streams, may trench adjacent portions of the two sheets. Offset gaps, instead of paired gaps, may thus date from the very inception of superposition.

But even were paired gaps, along either northwest- or southeast-flowing streams, formed at this early date, the conception of regional superposition from a surface of pre-Schooley date is distinctly favorable to the early abandonment of some of the gaps, with concurrent development of rectangular stream courses between the remaining offset gaps. As we have already seen, the depth of possible stream incision was so limited because of small vertical distances separating the Fall Zone peneplane

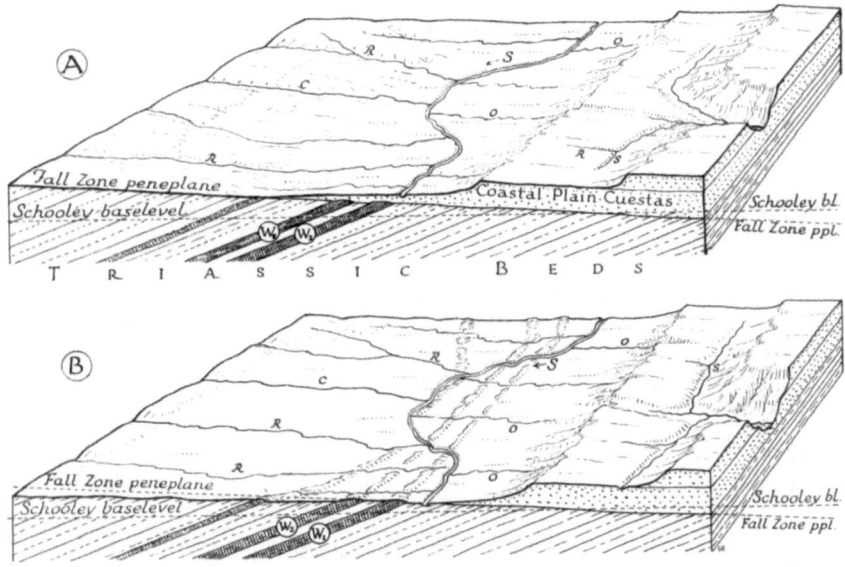

FIGURE 17. Inferred Cuesta and Lowland Topography above the Watchung Trap Sheets in Schooley Time

        C   Possible consequent stream flowing southeast
   R, R   Resequent streams flowing southeast
   O, O   Obsequent streams flowing northwest
   S, S   Subsequent streams draining broad lowlands
W_1, W_2 First and Second Watchung Mountain Trap Sheets

from the Schooley peneplane, and the Schooley peneplane from present baselevel, that extensive drainage modifications were not possible. But the long period of time during which the exposed trap sheets were slowly being reduced to Schooley baselevel, and the lapse of post-Schooley time, together represent an ample opportunity for such adjustments of stream courses as the limited depth of stream incision rendered possible. We should therefore expect to find indications of drainage modification in the direction of trellis or rectangular stream patterns, but should also expect to find that these changes are of local extent only. It is precisely this type of apparently distinct stream adjustments of strictly limited scope which we observe in the southern part of the Watchung crescent.

The Watchung trap sheets, like those of the Connecticut Valley, appear to be occasionally broken by oblique faults. Subsequent ravines developing along the fault zones may give oblique notches or cols resembling oblique wind gaps; or such ravines may, under the most favorable circumstances, divert streams along the fault zones to give water gaps traversing the trap ridges obliquely, some of which water gaps may, through later stream captures, become wind gaps. For this reason it is not always safe to accept oblique notches, with or without streams in them, as indications of superposition of originally oblique streams. This is of interest in the present connection, both because extent of apparent offset is in some cases increased by the oblique position of the notches, and because conquest of drainage by a subsequent stream working along a fault zone may conceivably be responsible for the development of a gap offset from its neighbor in the adjoining ridge.

Our attempts to restore an initial consequent drainage system in the southern part of the Watchung crescent which would, through normal readjustments, give the present distribution of stream courses, water gaps and wind gaps, have not been altogether successful. The principle that in a given ridge wind gaps should be progressively lower and lower when proceeding from the gap of the master stream in the direction in which the successive captures were effected, must be modified by the principle that in a given drainage basin the gaps of successive tributaries may initially be successively higher and higher as one gets farther from the master. Thus, if capture takes place with sufficient rapidity, successive wind gaps may still be found to be higher, instead of lower, as one proceeds in the direction of capture. The problem is further complicated by the fact that streams which formerly flowed through some of the gaps doubtless captured neighboring smaller streams, thereby producing one or more wind gaps, perhaps in a direction opposed to that in which later capture of the erstwhile pirate was accomplished. Add to this the further complications that some of the gaps were probably developed in the second or western ridge by streams which had shifted their courses before the first ridge was exposed; that in different places the gaps were cut by different types of streams flowing in opposite directions; that some gaps may represent the independent work of later erosion along fault zones; and that glaciation and impounded glacial waters may have changed the courses of streams and the depths of certain gaps; and it will be seen that the problem of restoring details of the original superposed consequent drainage of an earlier cycle must be nearly if not quite insoluble. Our efforts convince us that the simple succession of captures pictured by Davis will not explain the present topography of the Watchung region, and permit us to restore with some confidence parts of the earlier drainage pattern; but they indicate the impossibility of picturing with any assurance of accuracy the detailed processes by which each gap or set of gaps was formed. We believe that more than one type of drainage modification has left traces in the landscape, even if we cannot with certainty attribute each part of the resulting drainage pattern to its initial cause.

The Northwest-flowing Streams

Some of the considerations set forth in the preceding section suggest an interpretation of certain northwest-flowing Jersey streams which is quite different from the explanation advanced by Davis to account for their apparently abnormal courses. According to this author Ambrose's Brook, Manalapan and upper Assanpink Creeks (see Fig. 16) are new streams consequent on a recent slight deformation which tilted downward toward the northwest the latest peneplane (post-Schooley and supposedly Tertiary) of the region. Thus the direction of these streams, opposite to the inferred original slope of the peneplane, is attributed to a local reversal in slope of the old erosion surface as a result of warping.

Millstone River (Fig. 16), another northwest-flowing stream, is given a far more complicated history. The fact that it crosses directly through two fairly formidable barriers, the fall-line escarpment (described by Davis as a fault scarp indicating displacement of approximately one hundred feet) and the Rocky Hill trap ridge, apparently precluded interpretation of the Millstone as a simple consequent stream newly formed on the northwest-tilted peneplane. Accordingly Davis imagined that the original Millstone was a large river flowing from northwest to southeast through the very channel in which the present stream pursues an opposite course; and that this large river continued southeastward, past the present headwaters of the reversed stream, to the ocean.

It was then broken in two at the head of the present river where the greatest elevation of the central plain occurred, and thus had its former headwaters reversed from the southeast to a northwest direction of flow across and against the fall-line break by the tilting of the plain. Only in this way can the deep gap in Rocky Hill be explained. The river is thus consequent on the tilting of the plain, and yet antecedent to the accompanying faulting. It cannot be called an original stream, for it had an ancestor in its very channel. It is not a purely consequent stream, for it runs against the heaved side of a fault. It is not a strictly antecedent stream, for it flows in a direction determined by a disturbance that occurred late in its life. It is too exceptional a stream to have a generic name. We cannot expect to find many others like it.[11]

To Davis we owe our understanding of the fact that in the normal process of coastal plain dissection there develop cuestas upon the infacing slopes of which obsequent (opposite to consequent) streams flow in a direction opposite to that of the consequent drainage. As was noted in an earlier paragraph, obsequent streams on the New Jersey coastal plain should flow toward the northwest. Examination of New Jersey topographic maps reveals the fact that Manalapan and upper Assanpink Creeks, and the present headwaters of Millstone River, are not exceptional cases of northwest-flowing streams. They are members of a great family of streams occupying a continuous belt of territory from Raritan Bay to Delaware Bay, all of which head in the cuesta of the New Jersey coastal plain, and flow in parallel courses northwestward into the inner lowland (Fig. 16). They present all the characteristics of typical obsequent streams, and we need assume for them no complicated history. That the portion of the plain on which they flow has an inclination downward toward the northwest is a necessary consequence of the fact that this portion of the plain was carved by streams which from their inception flowed in that direction. Even were the floor of the lowland reduced to a peneplane, the surface of the peneplane must locally slope in the direction of the streams which carve it. That such peneplane slopes may be quite perceptible can be verified by any one who will examine portions of the Harrisburg peneplane adjacent to the Kittatinny and other Appalachian ridges.

Regional uplift will permit streams to trench a plain previously fashioned by them, and of course such uplift may be accompanied by warping. But in the present instance there seems to be no need to invoke warping to account either for the northwest courses of the streams or for the northwest slope of the plain down which they flow. The observed geomorphic relations appear to be altogether normal, and to indicate a normal and simple past history for the streams in question.

Ambrose's Brook and the lower portion of Millstone River (Fig. 16) belong in a somewhat different category from the streams discussed above. They are situated farther to the northwest, do not lie on the inface of a coastal plain cuesta, and appear to be exceptional streams rather than members of a great family showing identical characteristics. Of the two, Ambrose's Brook is the smaller and presents the simpler relations. It might, perhaps, be explained as in the nature of a resequent stream developing in the direction of dip of the Triassic strata.

The lower course of the Millstone likewise trends, in some measure at least, with the dip of Triassic beds; but its course athwart the two barriers bespeaks a less simple history. Of these barriers the Rocky Hill trap ridge requires no comment. Concerning the other, it should be pointed out that the early interpretation of the fall line as the expression of a fault with upthrow to the west no longer enjoys the confidence it commanded when Davis adopted it in his discussion. It is possible that in some places a fault may coincide with the feature known as the fall line; but it has been shown that this feature has independent existence as the line where two peneplane surfaces of different angles of slope mutually intersect. In the particular locality under discussion the rise against which the Millstone maintains its course might be interpreted as a true fault scarp, or as a fault-line scarp not affected by any recent displacement, the fracture in either case being quite unrelated to the true fall line, and perhaps located some distance from that line. Or it might be regarded as the steeper seaward slope of the Fall Zone peneplane, preserved on harder portions of the Triassic series and marking the location of the fall line, without genetic relation to any fracture. Theoretically it might even be nothing more significant than the outcropping of trap and certain Triassic sandstones locally more resistant than adjacent formations; for the rise is by no means so high, so continuous, or so clearly defined a feature as would be inferred from Davis' diagrammatic representation of it.[12] Map study of comparative elevations northwest and southeast of the supposed fall-line "scarp," a field inspection of the portion between New Brunswick and Princeton, and a preliminary profile study leave us in doubt as to the true nature of the poorly defined topographic break in the area under discussion. Further study of this feature is required.

Whatever be the true nature of the supposed fall-line barrier at the point where the Millstone River crosses it, we have to explain the course of the stream transverse to this obstacle and to that represented by the Rocky Hill trap ridge. It has occurred to me that a comparatively simple explanation might be found if we regard the lower Millstone (and possibly also Ambrose's Brook) as a superposed obsequent stream, the northwest course having been inherited from a time when the cuesta of the coastal plain was farther northwest than now, although not so far northwest as indicated in Figure 17. Such an interpretation is frankly speculative, and not wholly convincing in view of the markedly oblique course of the lower Millstone, discussed under the next heading; but it seems less radical, and far less complicated, than the history suggested by Davis. Opposed to that history are the theoretical objections to the process of completely reversing a large stream in its channel, already set forth in our discussion of the supposed reversal of Pennsylvania drainage; and the practical objections that the tributaries of the Millstone have a pattern inconsistent with a former southeast direction of flow along the main channel, while the topography of the cuesta fails to show any evidence that it was ever crossed by a relatively large southeast-flowing stream in this region. When we consider the recency of the supposed reversal, and the comparative elevations of the two barriers, the cuesta, and the intervening country, it would seem that any former southeast drainage must have left a record of its existence which the moderate changes of later time could not have effaced.*

The evidence rather suggests a northwest drainage of very early origin, still maintained on the present cuesta front, but largely destroyed by subsequent erosion farther northwest where only an occasional superposed obsequent is preserved, sometimes traversing hard rock barriers discovered through dissection of the underlying structures. On the assumption that the true fall-line angle lies southeast of Princeton, this interpretation implies that an occasional obsequent stream could maintain its northwest course superposed on the underlying Triassic rocks, while these and the nearby coastal plain series were reduced to the Schooley level, as well as during the shorter period of post-Schooley dissection. Such a possibility is favored by the fact that the vertical distances through which streams of this region could intrench their channels were necessarily small, for reasons explained in an earlier paragraph; and by the fact that the Millstone River lies halfway between the nearest southeast-flowing consequents, the Delaware and the Raritan. The first consideration means that deflection of a superposed obsequent through capture would in this region be comparatively difficult; the second means that the Millstone River would be one of the most difficult obsequent streams for subsequent branches of southeast-flowing consequents to reach, and hence would be one of those most likely to persist along its original path into the present cycle.

* Temporary local reversal of the Millstone current by one or more ice advances which may have blocked the lower part of the stream is not excluded. Boulders of trap reported from a gravel bed at Kingston, *upstream* from the adjacent trap ridge, may point to such a history, although we must consider the possibility that they were brought *downstream* from parts of Stony Brook heading on the trap ridge farther southeast. Such temporary reversal would not turn the stream through the high ground of the coastal plain cuesta, but southwestward along the lowland to the Delaware.

REFERENCES

[1] William Morris Davis: "The Rivers of Northern New Jersey, with Notes on the Classification of Rivers in General," *Nat. Geog. Mag.*, Vol. 2, 81–110, 1890. *Geographical Essays*, 485–573, Boston, 1909. See also William Morris Davis and J. Walter Wood, Jr.: "The Geographic Development of Northern New Jersey," *Bost. Soc. Nat. Hist., Proc.*, 24, 365–423, 1889.

[2] William Morris Davis: "The Rivers of Northern New Jersey, with Notes on the Classification of Rivers in General," *Nat. Geog. Mag.*, Vol. 2, 81–110, 1890, see p. 92.

[3] *Ibid.*, see p. 91.

[4] William Morris Davis and J. Walter Wood, Jr.: "The Geographic Development of Northern New Jersey," *Bost. Soc. Nat. Hist., Proc.*, 24, 365–423, 1889, see p. 387.

[5] J. Volney Lewis: "The Double Crest of Second Watchung Mountain," *Jour. Geol.*, Vol. 15, 39–45, 1907.

[6] Rollin D. Salisbury: "The Physical Geography of New Jersey," *Geol. Surv. N. J., Final Report of State Geologist*, Vol. 4, 1–170, 1898, see pp. 86–87, 92–93.

[7] Rollin D. Salisbury: "Surface Geology — Report of Progress," *Geol. Surv. N. J., Ann. Rept. State Geologist for 1893*, 47–48, 1894.

[8] Henry B. Kümmel: Personal communication.

[9] William Morris Davis and J. Walter Wood, Jr.: "The Geographic Development of Northern New Jersey," *Bost. Soc. Nat. Hist., Proc.*, 24, 365–423, 1889, see pp. 390, 401.

[10] William Morris Davis: "The Rivers of Northern New Jersey, with Notes on the Classification of Rivers in General," *Nat. Geog. Mag.*, Vol. 2, 81–110, 1890, see p. 99.

[11] *Ibid.*, see p. 108.

[12] *Ibid.*, see Fig. 1.

CHAPTER SEVEN

SUPERPOSED SUBSEQUENT DRAINAGE OF THE WATCHUNG CRESCENT

It may have occurred to the reader that the theory of superposition of obsequent streams, set forth in the preceding section of this volume, involves as a corollary the superposition of one or more subsequent streams to which the obsequents were tributary, as shown in Figure 17. The lower Millstone may itself belong to the class of superposed subsequents; for its oblique course, midway between that of a typical obsequent (see preceding section) and a typical subsequent, leaves us in doubt as to its true origin. It will be noted (Fig. 16) that where the upper Millstone joins Stony Brook southeast of Princeton, the Stony Brook-lower Millstone combination gives us a stream flowing in the manner of a subsequent, nearly at right angles to the trend of the upper Millstone, which is undoubtedly obsequent. But the case is not sufficiently clear, especially since the lower Millstone as a whole takes a course in general obliquely across the weak rock belt of the Triassic. It is therefore pertinent to inquire whether elsewhere along this belt of country there are trustworthy indications of superposed subsequent drainage.

The Davisian Interpretation

In his interpretation of New Jersey drainage Davis explained the gaps through the Watchung Mountains at Paterson and Millburn * (or Summit) as the work of two major superposed

---

* We follow the local spelling and that employed on maps of the New Jersey topographic survey. In Davis' text and on the United States topographic quadrangle the name appears in the form "Milburn."

consequent streams, the Pequannock-Passaic and the Rockaway-Rahway (Fig. 18). A study of these gaps indicates that any satisfactory explanation of their history must account for the following pertinent facts: (1) the gaps are remarkably broad when compared with other gaps in the ridges cut by transverse southeast-flowing streams; (2) the broad rock floor of the Paterson gaps (exclusive of the narrow trench of probable post-glacial age) appears to be distinctly higher than the rock floor of the Millburn gaps; (3) the paired gaps at Paterson are aligned in a northeast-southwest direction; (4) this trend is continued toward the southwest in the broad gap between Riker Hill and Hook Mountain; (5) there is a marked contrast between the two sections of the Watchung ridges lying north and south of Millburn gap in respect to the number of water gaps and the disposition of wind gaps cut in the ridges.

If we examine the Davisian interpretation of Watchung drainage with these facts in mind, we observe:

(1) At Paterson and Millburn larger gaps than elsewhere are to be expected, although the degree of contrast actually observed seems disproportionate to the presumably moderate contrast in size between these and other southeast-flowing streams, all of which were comparatively small and the largest of which were believed to head in the highlands but a few miles farther west.

(2) The difference in levels of the rock floors of the two principal gaps is opposed to the Davisian interpretation, by which the Millburn gap is supposed to have been robbed of its transverse stream some time ago, and according to which this gap should therefore have a higher floor. Glacial overdeepening of Millburn gap will hardly explain a difference which appears to be marked, especially when we consider the nearness of the gap to the limit of glaciation, and the probable direction of ice advance. The possible effect of warping in reversing the relative elevations of the gaps is discussed in a later paragraph. Here we need only observe that the amount of warping determined

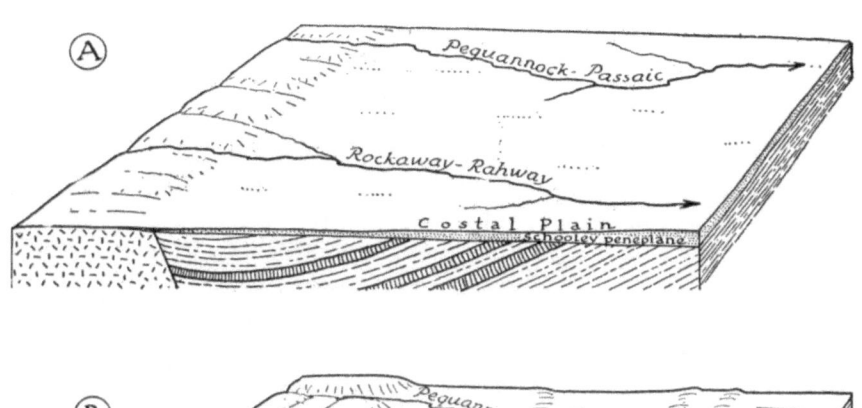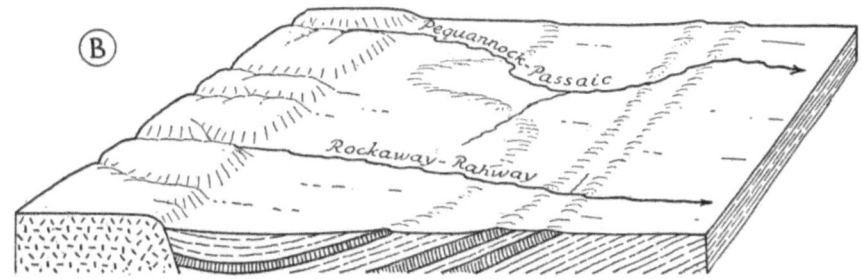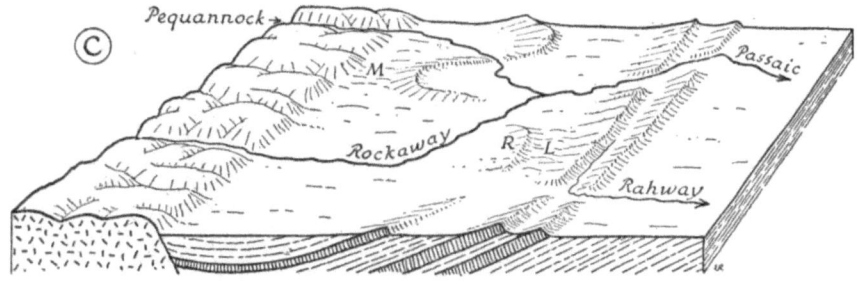

FIGURE 18

Superposition and Adjustment of Watchung Drainage, According to Davis

for this region is insufficient in itself to account for the observed relations.

That the depth of the Millburn gap is distinctly greater than that at Paterson is indicated by our field inspection of the two areas, and is confirmed by data supplied by Kümmel, cited below. Salisbury [1] realized this fact, and seems to have inferred that the larger stream passed through Millburn gap (Fig. 19). He does not specifically recognize any diversion from either

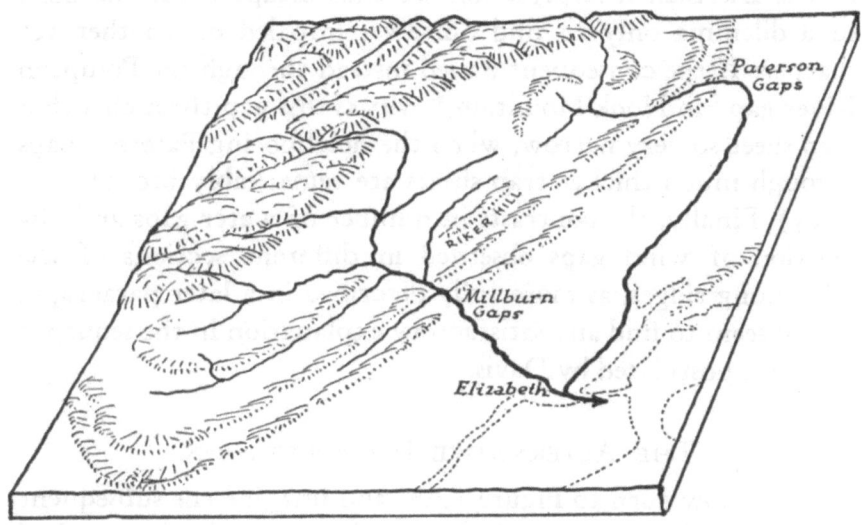

FIGURE 19

Diagram Showing the Supposed Course of Drainage in the Watchung Region Previous to the Last Glacial Invasion, According to Salisbury

gap by capture, but otherwise pictures a drainage pattern similar in its broad outlines to that described by Davis. If Salisbury's analysis of the drainage history explains the observed difference in levels of the Paterson and Millburn gaps, and avoids the necessity of a capture which we shall find difficult to envisage, it offers no other advantage over the Davisian interpretation.

(3) The alignment of the two gaps at Paterson is in a direction clearly abnormal, if regarded as the work of a super-

posed consequent stream incised from a southeast-sloping coastal plain surface.

(4) This abnormality is increased if we regard the gap north of Riker Hill as due to erosion by the superposed consequent stream. Davis apparently would not carry his consequent master stream so far toward the southwest, and he may have believed that local thinning out and disappearance of the trap sheet was responsible for this particular gap. But on the basis of the Davisian interpretation we thus escape from one horn of a dilemma only to find ourselves impaled on another yet sharper: If the consequent passed instead through the Pompton River gap [2] in Hook Mountain,* why is this gap through a thin trap sheet so very narrow, when the neighboring Paterson gaps through much thicker trap sheets are remarkably broad?

(5) Finally, the contrasts in number of water gaps and disposition of wind gaps observed in different sections of the Watchung ridges, as more fully described in a later paragraph, do not seem to find any satisfactory explanation in the sequence of events postulated by Davis.

The Alternative Interpretation

Let us now turn to Figure 17 A, and imagine the subsequent stream (S) flowing toward us from the middle background of the diagram, to be superposed upon the trap sheets visible in the front of the diagram, in such manner as to give, at a later stage, the conditions represented in Figure 17 B. What are the expectable consequences of such a history?

Great breadth of the gaps. — Since a subsequent stream like that inferred could, in New Jersey, gather drainage from considerable areas of Triassic rocks to the northeast and crystalline rocks to the north and northwest, it might well be relatively large as compared with other streams traversing the ridges of the Watchung crescent. Possibly the Ramapo earlier pursued

---

* This is the position assigned to it by Davis and Wood. See Figure 18, C, northeast of M.

a course outside the crescent, to contribute its waters to the subsequent before the latter reached the Paterson gaps. As the evidence in favor of a large subsequent stream developed we early looked to the Hudson River as a possible source of considerable volume. If this stream formerly turned southwestward across the Palisades trap ridge (Fig. 21) to become the subsequent in question, the relations would, on the preëxisting coastal plain cover, resemble those exhibited by the Delaware and Susquehanna where these turn southwestward, in part at least along the inner lowland of the present coastal plain. Such history calls for a very broad gap in the Palisades ridge the rock floor of which should be slightly higher than the Paterson gap (160-180 feet) in First Watchung Mountain. Examination of maps showed that such a gap, strikingly different from all other gaps in the Palisades, exists at Sparkill (Figure 16), 15 miles northeast of Paterson and so in a position normal for the supposed subsequent river. According to the maps the Sparkill gap is two miles broad, and has its bottom (exclusive of an inner gorge similar to that at Paterson) from 180 to 200 feet above sea level. It will thus be seen that Sparkill gap compares favorably in size and elevation with the other gaps supposed to have been cut by the large subsequent river. There is much glacial debris in Sparkill gap, but Berkey[3] has mapped numerous outcrops of trap in place up to elevations slightly above the 200-foot contour, while our field examination confirms the fact that the rock floor of the gap varies in elevation from about 180 to slightly more than 200 feet. In general appearance the gap quite duplicates those at Paterson and Millburn. Thus there is found strong presumptive evidence that the earlier Hudson River [*] was the subsequent with which we have to deal in the Watchung area.

In any case, we can safely say that no observed topographic

[*] The reader who examines the Sparkill gap on the Tarrytown topographic quadrangle will of course remember that the Hudson River referred to is not the broad sheet of water occupying the present drowned valley, but the stream of moderate volume which was found here before its valley was submerged.

relation definitely excludes the possible presence of a relatively large superposed subsequent stream in the Watchung region in pre-glacial time, while certain features strongly indicate that such a stream must have existed. It is evident that a large subsequent stream should carve appropriately large gaps where it crossed and recrossed the Watchung ridges. Nor is this the only significant consequence of size. If the stream were large, and the vertical distance available for intrenchment comparatively small, the superposed course would long be maintained. It might easily persist not only through the Schooley cycle, but also far into post-Schooley time, thus cutting the gaps well down toward present sea level.

Relative levels of Paterson and Millburn gaps. — The Paterson gap, being farther upstream than the Millburn gap, should be the higher of the two, unless relative levels were altered by glaciation or by warping, an event regarded as improbable for reasons already stated. This difference in levels would be accentuated if the upper waters of the superposed subsequent stream were first diverted from Paterson gap or Sparkill gap, or from both, to courses more favorably situated on weak Triassic beds east of the Watchungs and Palisades, while the lower waters, including all the drainage gathered within the crescent, continued for a time to flow through Millburn gap.

Oblique alignment of gaps near Paterson. — The northeast-southwest alignment of the Paterson gaps, and the extension of this alignment to the gap north of Riker Hill, find reasonable explanation in the general northeast-southwest course of the subsequent, complicated by the moderate swinging expectable in a fairly large river on the floor of a broad lowland developed in soft beds of the overlying coastal plain. The problem of accounting for the Paterson gaps was presented for discussion to my seminar in physiography, and among the solutions offered was one by my research assistant, Miss Clara Rom, invoking superposition of a large stream flowing nearly with the strike of the Triassic beds but wandering moderately on its flood plain. As this interpretation was reached independently, we may perhaps find in that fact some confirmation of the reasonableness of the conception here set forth.

Direction of flow of the superposed subsequent drainage. — We have considered the possibility that this large subsequent stream may have flowed from southwest to northeast, to join the Hudson, rather than from northeast to southwest. But if present comparative elevations of the rock floor in different gaps are any indication of pre-glacial conditions, the former river must have pursued a course toward the southwest. The figures on page 112, arranged in descending order along the supposed course of the large subsequent river, are significant.

It will be observed that the available data apparently are distinctly favorable both to the conception that a large superposed subsequent stream crossed and recrossed the trap ridges, and to the idea that the direction of flow of this stream was from northeast to southwest. One should not, however, assign too much importance to the particular figures cited for the gap elevations. In the first place they are in some cases only approximate, in others incomplete. We have no satisfactory data for the comparative levels of the rock floor in the two gaps at Paterson. From figures given by Salisbury and Kümmel[4] one would infer that the trap occurs at a higher elevation in the gap through Second Mountain than it does in the First Mountain gap, although these authors discuss the particular elevations at which the present Passaic River encountered bedrock when it cut through overlying glacial debris, rather than the comparative elevations of the rock floors of the gaps as a whole. Were the glacial debris as fully removed from the Second Mountain gap as it is from the one just east, we might find the bottom of the western gap lower than the eastern; or careful leveling might show the reverse to be true. The latter result would be unfavorable to the idea that the large superposed subsequent flowed from northeast to southwest, although it would not necessarily disprove the validity of that conception. It must

Gap	Local conditions	Elevation of bedrock floor in feet above sea level
Sparkill Gap	Partly filled with glacial debris	180–200*
Paterson gap in First Mountain	Bedrock well exposed, partly due to artificial removal of glacial debris	160–180*
Paterson gap in Second Mountain	Deep covering of glacial debris, but bedrock exposed in certain localities	180, 170, less than 145, 165–172†
Gap north of Riker Hill	Floor completely buried under glacial and other deposits	Unknown
Gap south of Riker Hill	Floor completely buried under glacial and other deposits	106, 77, 113, 127, 141, 125, 132, 141, 137†
Millburn gap in Second Mountain	Floor deeply buried under terminal moraine deposits	Unknown
Millburn gap in First Mountain	Floor buried under glacial deposits	0, 0, 113†

* Based on field comparison of topography and rock outcrops with contour map, hence not precise.

† Data from field maps and well records furnished by Henry B. Kümmel, and presented in approximate order from southwest to northeast across each gap. Lowest figures may indicate gorge in floor of each gap, similar to gorges observed in floors of the two Paterson gaps and the Sparkill gap. 0 = rock encountered at sea level. Concerning a small exposure of trap in Millburn gap in First Watchung Mountain at elevation 144 feet above sea level, Kümmel writes: "If it is a large boulder it has, of course, no significance.... If it is part of the bedrock it is difficult to reconcile it with the testimony of the wells unless we assume that there is a very deep inner gorge in the Millburn gap." Interpretation of the outcrop as that of a glacial boulder accords best with all the evidence, even if we admit the existence of an inner gorge in the floor of the gap.

be remembered that not only in post-glacial time, but possibly also in interglacial time, a large drainage discharged eastward through these gaps, and this would tend to reduce the level of the eastern gap below that of the western, if the pre-glacial levels were in the opposite sense. Differential glacial erosion might reduce the eastern gap more than the western, although the direction of ice advance, in Wisconsin time at least, was not favorable to producing any great changes in the direction noted.

A second reason for using caution in accepting any particular figures for gap levels is the fact that the Watchung region is known to have suffered differential warping in post-glacial time. Kümmel [5] has determined the direction and amount of this warping, and from his studies it appears that the more northerly gaps have, in the period indicated, been raised higher than they were when the ice was present in the region. This does not mean, however, that the northern gaps are now higher than they were in *pre-glacial* time; and it is this comparison alone which is significant for us. It is distinctly stated that the land " was weighted down beneath the load " of the advancing ice; then following removal of the ice, " the land tended to regain something of its original position." It is the last movement alone which Kümmel could measure. If it exactly equaled the original downweighting of the land, then the present gap elevations are the same as those of pre-glacial time, and would, taken in their entirety, seem to demonstrate the southwestward flow of the superposed subsequent river which carved the gaps. But it is not likely that the land returned to precisely the same level it earlier maintained; hence the figures cited presumably contain an element of error due to warping. The probable error is not large, and can scarcely affect our general conclusion. Even if we subtracted the total known warping from the figures given, the result apparently leaves all, or nearly all, of the gaps in the same *relative* positions with respect to each other that they actually occupy.

One of my graduate students, Miss Mabel Schwartz, has directed my attention to the fact that the distribution of steep and gentle slopes where First and Second Watchung Mountains are cut by the Paterson gaps can better be explained as the work of a stream flowing southwestward into the crescent than as that of a stream flowing out of the crescent toward the east or northeast. While this consideration is favorable to the conception of a southwest-flowing Hudson River, we do not regard the direction of flow as a vital part of the theory of superposed subsequent drainage.

Regarding the broad gaps north and south of Riker Hill, let us note in passing that the discovery of trap in repeated well drillings across the southernmost of these two gaps effectively disposes of the possibility that this gap is due to thinning out and disappearance of the trap sheet, and creates two strong presumptions: (1) that the gap must have been carved by a river of large volume, and (2) that the similar gap north of Riker Hill must have had the same origin. The two broad gaps in the third trap ridge thus become important links in the chain of evidence tending to prove early superposition of a large subsequent river across all three ridges.

Objections to the Capture Theory

In this connection we must point out a further serious objection to accepting any such history of major stream captures as that pictured by Davis in the Watchung area. In describing these captures in his paper on " The Rivers of Northern New Jersey " Davis uses ideal diagrams which show but two of the three trap ridges of the district. The deductions based on ideal cases are then applied to the actual region in the words of a text which leaves us in doubt on certain critically important points. He tells us, for example, that the upper Passaic was the growing subsequent which captured the headwaters of the Rockaway-Rahway and other transverse streams farther to the southwest; but he does not discuss the relation of this subsequent to the third trap ridge (Long Hill-Riker Hill-Hook Mountain), omitted from his diagrams. In his earlier paper on the "Geographic Development of Northern New Jersey," published in collaboration with J. Walter Wood, Jr., the ideal diagrams do not appear, but a rough sketch map and the accompanying text represent the subsequent as lying west of the third ridge in the southern part of the area, and east of it farther north. We are not certain that Davis would have adhered to this interpretation at the time of writing his later paper, but in trying to portray his conception of the drainage history by diagrams which represent all three trap ridges (Fig. 18), I have been compelled to give the pirate subsequent some location with respect to the third ridge, and have chosen the one sketched in the earlier paper cited above. We are not at liberty to take the present position of the upper Passaic as necessarily the original position of the piratical subsequent for, as Davis points out, glaciation has materially shifted stream courses within the Watchung crescent. It is properly within the limits of the Davisian interpretation to give to the conquering stream the most favorable position possible, whether or not it be the particular one chosen for representation in his earlier paper.

Now it is precisely when we seek a favorable location for the pirate with respect to the third ridge that the theory of capture breaks down. There is no favorable location. We may place it between the second and third ridges in the southern part of the crescent, where the upper Passaic now flows; but we cannot carry it northward to Paterson gap along this subsequent valley, because a large area of uneroded Triassic rocks, much higher than the floor of Millburn gap, blocks this route. At Livingston (L, Fig. 18, C) midway between the second and third ridges, a well of the Livingston Sand and Gravel Company starting at a surface elevation of 328 feet above sea level passed through 94 feet of glacial deposits, then struck bedrock at elevation of 234 feet. It penetrated 27 feet of red shale before drilling ceased. Examination of the local topography convinces us that in the narrow space between the two ridges there is not room, in view of the known high elevation of bedrock in the middle of the depression, for a subsequent stream to pass this route at an elevation sufficiently low to tap a large stream passing through Millburn gap at elevation of 113 feet or less. This conclusion holds, even if we assume that the known amount of post-glacial warping is a fair measure of differences between relative levels of pre-glacial and post-glacial time (see discussion of this question on an earlier page).

If we place the captor back of (west of) the third ridge, we are no better off, for then we must carry it through the narrow pass at Montville (M, Fig. 18, C) where Hook Mountain swings westward close to the crystalline highlands. The rock walls of this pass are occasionally exposed at elevations of approximately 260 and 280 feet where the depression is only a mile wide, while bedrock outcrops on the floor close to the 200-foot contour a third of a mile from the east wall. Again the topography seems to preclude the possibility of a piratical subsequent's passing by this route at an elevation sufficiently low to capture a transverse stream flowing through Millburn gap.

There remains alone some course across the third ridge. That a small subsequent brook should be able to gnaw headward through a resistant trap ridge and capture a large stream lying beyond it, is far from probable, and is not envisaged in the drainage history portrayed by Davis. Yet the position of the third ridge makes capture impossible without this improbable event, unless we accept the almost equally improbable alternative that the gap north of Riker Hill (R, Fig. 18, C), unlike the one south of that hill, was never occupied by trap.* Thus we are again impelled to turn to the theory of superposed subse-

---

* The general continuity of all three trap sheets, proved in most cases by surface exposures or by well records, makes one hesitate to infer local thinning out and disappearance of the trap at this particular locality. The apparent continuity of direction of the curved crest of Riker Hill with that of Hook Mountain, renders disappearance of the trap due to faulting an improbable interpretation.

quent drainage for a simple and rational explanation of all the observed facts.

Significance of Minor Water Gaps and Wind Gaps

The appeal to a large subsequent stream developed on coastal plain beds and involving a coastal plain topography both of which have completely disappeared from the region under discussion may impress the reader as somewhat too daring. Yet no conditions have been implied, and no processes invoked, which are not normally, and even almost necessarily, inherent in the probable history of the region. The great size of the major gaps in the three trap ridges, the position of these gaps in respect to one another, and the relative levels of their floors, all point to the correctness of the interpretation here offered. But there remains a more delicate test of this theory of superposed subsequent drainage than any yet applied. When the study reached the point where it became evident that only by postulating a large river lying partly within and partly without the Watchung crescent could we account satisfactorily and simply for all of the observed facts, it occurred to the writer that a necessary corollary of this hypothesis must be a distinct contrast in the number of minor water gaps and in the disposition of minor wind gaps along the two sectors of the Watchung ridges north and south of the major Millburn gap. For a full understanding of this point some preliminary discussion is necessary.

If we turn again to Figure 17, we note that aside from the major subsequent river two types of streams will be superposed across the trap ridges. Where the large subsequent river lies east (to the right) of the ridges, the latter are crossed by comparatively long superposed consequent (C) or resequent (RR) streams. Where the large subsequent lies west of the ridges, the latter are traversed in the opposite direction by relatively short obsequent (O, O) streams. Other things being equal, it would seem that the longer (and larger) consequents and resequents should resist capture for a greater time than would the shorter (and smaller) obsequents. Accordingly we might expect to find today a larger number of water gaps along those sections of the trap ridges where the superposed subsequent lay east of the ridges; and a smaller number, if any, where the subsequent lay west of the ridges.

Where the transverse streams have been diverted to more favorable courses along weak rock belts wind gaps may remain in the hard trap ridges as evidence of the former presence of the streams. Let us consider the conditions under which these wind gaps are formed. It is obvious that so long as the superposed subsequent occupies the position shown in Figure 17 B, the superposed consequents and resequents, flowing eastward across the ridges to join the subsequent where it lies east of the ridges, must have higher water gaps in Second Watchung Mountain and lower water gaps in First Watchung Mountain, since the latter ridge is always crossed farther down the courses of the transverse streams. Conversely, the superposed obsequents will, for similar reasons, have higher gaps in First Watchung Mountain. Whether these relative levels will persist unchanged after the transverse streams have been diverted and the water gaps transformed into wind gaps, will depend on the conditions under which capture is effected.

If a transverse stream, DB (Fig. 20, A), is captured by a subsequent, S, the relative levels of the gaps 1 and 2 will normally remain unchanged thereafter. So long as the shrunken remnant of the beheaded stream, B (Fig. 20, B) continues to flow through both gaps, the relative levels cannot be reversed. If a tributary, T, of the diverted stream, D, pushes the divide back to the floor of gap 2, leaving the headwaters of B flowing through gap 1 alone, the level of the latter gap may be further lowered, thus accentuating the difference of levels already existing. Only under exceptional circumstances could the tributary, T, ever push the divide beyond the hard rock ridge in which

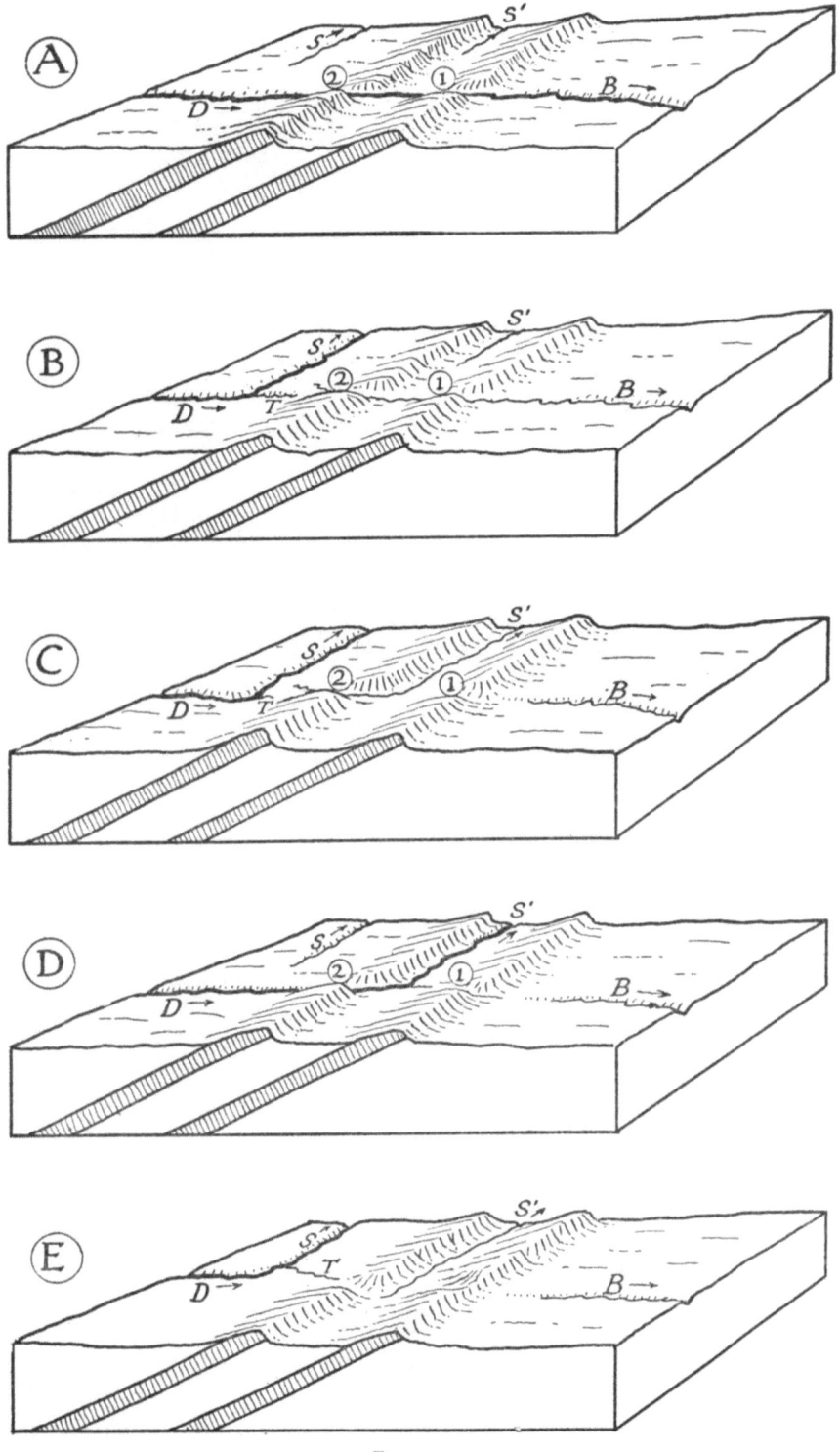

FIGURE 20

Development of Paired Wind Gaps

gap 2 is located. But a second subsequent, S', gnawing headward between the two ridges, may in time bring about a further diversion of the upper waters of the already beheaded stream, B (Fig. 20, C). This may occur under conditions which will leave a small stream still flowing through gap 2, while gap 1 is left as a deserted wind gap. But even under these circumstances the cases will be rare in which deepening of gap 2 continues to such an extent that it becomes lower than gap 1; for capture by a subsequent located between two closely adjacent ridges will, for reasons already discussed, be long delayed, and such delay will afford the tributary, T, opportunity to push the divide back to the floor of gap 2 and thus assure its early conversion into a wind gap.

The history will be different if the first capture is effected by the subsequent, S', flowing between the resistant ridges, to give the conditions represented in Figure 20, D. Obviously in this case the large stream, DS', may continue to cut gap 2 for a long time after gap 1 has been converted into a wind gap, and thus the level of gap 2 may be reduced well below that of gap 1. If now the upper portion of the stream, DS', be captured by the subsequent, S (Fig. 20, E), we shall have two wind gaps in which the original relation of levels has been reversed. But, while this is a possibility to be considered, it is not a probability to be expected. As we noted when studying the degree of drainage adjustment along the Watchung ridges, the development of subsequent streams along a narrow belt of weak rock between closely adjacent hard rock ridges proceeds slowly and with great difficulty. Streams on broader belts of weak rock back of the ridges, such as S, enjoy distinct advantages over streams located between the ridges, and will ordinarily divert the transverse drainage long before streams like S' can accomplish that feat. In other words, where two closely adjacent hard rock ridges were once crossed by superposed transverse drainage, the wind gaps which afford evidence of that drainage will normally preserve their original relative levels. This would mean that for a region like that shown in Figure 17, B, diversion of the superposed consequents and resequents crossing the ridges in the front and rear portions of the diagram should leave a series of wind gaps in ridge W_1 *lower* than the corresponding wind gaps in ridge W_2; whereas diversion of the superposed obsequents crossing the ridges in the middle ground of the diagram should leave a series of wind gaps in ridge W_1 *higher* than the corresponding wind gaps in ridge W_2.

Test of the Theory of Superposed Subsequent Drainage

We may apply the principles based on the foregoing deductions as a test of the validity of the theory that the Paterson and Millburn gaps were carved by a master subsequent stream superposed on the trap ridges in such manner that it flowed into the Watchung crescent at Paterson and out again at Millburn, essentially as represented in Figure 17, B. In doing this we encounter at the outset certain difficulties. While there are three trap ridges involved in the Watchung crescent, the western ridge is so far removed from the others in many places as to expose a very broad belt of weak rock between them. Since our analysis of wind gap levels applies only in those cases where the resistant ridges are closely spaced, we cannot include in our test the third or westernmost ridge making up Long Hill, Riker Hill, and Hook Mountain. Fortunately there remain two closely spaced ridges, First and Second Watchung Mountains, to which the test is properly applicable.

A further difficulty arises from the fact that north of the Paterson gaps First Mountain is represented by a short and imperfectly developed segment, while the northward extension of Second Mountain presents evidence suggesting that some, at least, of its gaps have been altered in glacial times, probably in part by glacial waters. We cannot, therefore, expect to secure reliable evidence of former stream directions from this part of the area. But again, fortunately, we have two long sections

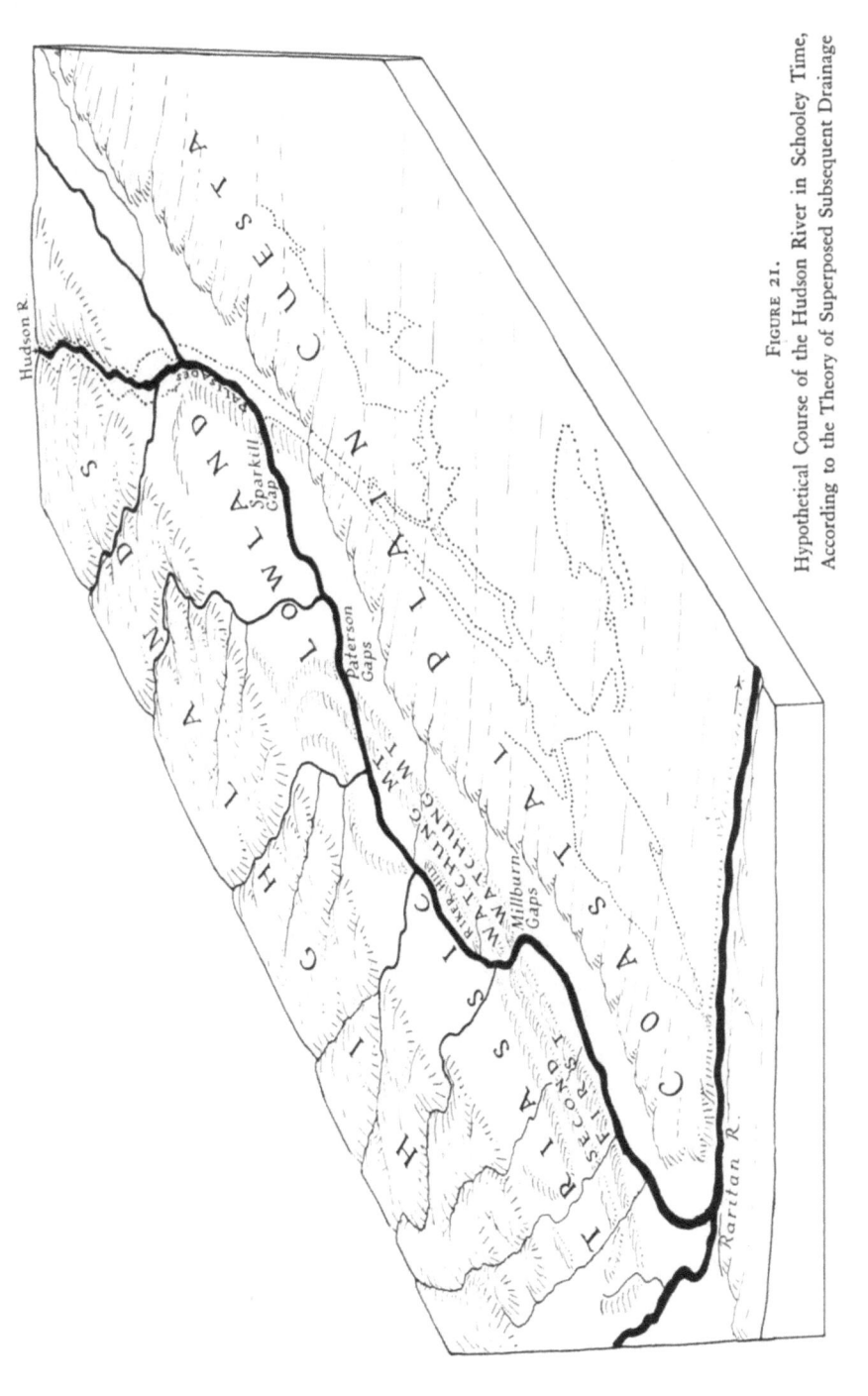

FIGURE 21.
Hypothetical Course of the Hudson River in Schooley Time, According to the Theory of Superposed Subsequent Drainage

of the trap ridges which are normally developed and apparently not seriously affected by disturbing influences, and which are accordingly properly comparable.

North of Millburn gap is a section fourteen miles long, where according to our theory the trap ridges were traversed by superposed obsequent streams, presumably of limited length and moderate volume, flowing northwestward to join that part of the master subsequent river located within the crescent, *i.e.*, west of the ridges (middle ground of Figure 17, B). If our theory of superposed drainage and our analysis of the test of its validity be correct, we should find few water gaps along this section, and the higher wind gaps should be found along the crest of First Mountain.

South of Millburn gap and extending to where the crescent makes a marked bend toward the west (Fig. 16), is a section of almost exactly the same length (fifteen miles) where, under our interpretation the trap ridges were crossed by superposed consequent and (or) resequent streams, some at least presumably of appreciable length and considerable volume, flowing southeastward to join the master subsequent east of the ridges (foreground of Figure 17, B). Here, if our reasoning be correct, we should find a larger number of water gaps than in the first section, while the higher wind gaps should notch the crest of Second Mountain.

What are the facts? In the section north of Millburn gap the First Watchung Mountain is not crossed by a single water gap, although at Great Notch south of Paterson, where this ridge is double, a tiny stream holds its course across one half of the double ridge. Likewise in Second Watchung Mountain not a single water gap is encountered. In contrast with this situation in the north, we find in the section south of Millburn gap 2 water gaps (or 3 if we include Chimney Rock gap at the south end of the section) cutting through First Mountain; while 4 water gaps are cut in the crest of Second Mountain, although none of the 4 still drains areas as far west as the lower backslope of the ridge. It seems fair to state the comparison as 7 water gaps in the southern section against none, or but a partial one, in the northern section. The testimony of the water gaps seems to confirm the validity of the drainage history set forth above.

The evidence afforded by the wind gaps is even more convincing. In the section north of Millburn gap there are 7 wind gaps in First Mountain which are *higher* than gaps opposite, or nearly opposite, them in Second Mountain; and no clear case of a gap in First Mountain which is lower than the corresponding gap in Second Mountain. South of Millburn gap the relations are completely reversed. Here we find 6, or possibly 7, wind gaps in First Mountain which are *lower* than the gaps behind them in Second Mountain; and no wind gap in First Mountain which is clearly higher than the corresponding gap in Second Mountain. There are a few doubtful cases where paired wind gaps are located within the same contour interval (that is, the vertical range of their respective elevations cannot exceed twenty feet) but precise elevations are lacking; or where one may be uncertain as to which gaps in the two ridges should be paired. But the evidence as a whole is clear and unmistakable to this effect: that north of Millburn the wind gaps are prevailingly high in First Mountain and low in Second Mountain, indicating a northwestward flow for the superposed streams; while south of Millburn the wind gaps are prevailingly high in Second Mountain and low in First Mountain, indicating a southeastward flow for the superposed streams.

When the hypothesis of a large superposed subsequent stream in the Watchung region was first entertained, it was tentatively invoked to account for the large size of the Paterson and Millburn gaps, the oblique position of the Paterson gaps, and the greater depth of the Millburn gap. The apparently greater depth of the rock floor of the western gap (in Second Mountain) as compared with the eastern gap (in First Mountain) at Paterson, and the fact that the rock floor in the gap south of

Riker Hill is considerably lower than in either of the above, were later discoveries which seemed to support the hypothesis. But not until the former drainage relations had been worked out in some detail did it occur to the writer that an examination of water gaps and wind gaps in two contrasted sections of the trap ridges should afford evidence tending to confirm or deny the validity of the hypothesis. The greater abundance of water gaps in the southern section had previously been noted as a fact not satisfactorily explained by the Davisian interpretation; but the contrasted relation of wind gap levels was not suspected until sought for in an effort to test the hypothesis. The reader may therefore assign this evidence whatever weight seems appropriate to the discovery of new facts giving unexpected confirmation of a hypothesis previously elaborated.

Possible Southwestward Extension of the Master Subsequent

Salisbury,[6] like Davis, made the Millstone (Fig. 16) from its junction with Stony Brook down to the Raritan a reversed stream; but he considered that the former southeast-flowing river turned southwest at a point near Princeton, to enter the Delaware. Now it will at once be apparent that if the superposed subsequent which carved the gaps at Paterson and Millburn continued southwest to join the Delaware, instead of contributing its waters to a master southeast-flowing stream in the position of the Raritan, Salisbury's interpretation could be brought into harmony with the ideas expressed in these pages. This would be true not only with respect to our conception of superposed southwest-flowing drainage in the Watchung area, but also as regards the suggestion offered in an earlier paragraph that the lower Millstone may belong to the class of superposed subsequents, although, like the superposed subsequent of the Watchung area, it was in part let down upon the underlying Triassic sandstones and trap while flowing in a direction oblique to these beds. The drainage pattern which results from prolonging the large subsequent of the Watchung area (Fig. 17, B) southwestward up the lower Millstone and Stony Brook (Fig. 16) to Port Mercer, thence southwestward over a low divide and along lower Assanpink Creek to the Delaware, is quite appropriate for a superposed subsequent river.

Against this hypothesis may be urged (1) the objections to reversing a large river in its channel, already set forth; and (2) the fact that the gap through which the lower Millstone crosses Rocky Hill trap ridge is much narrower than the Paterson and Millburn gaps. As regards the first point, the objections to reversing the lower Millstone are certainly less formidable than are those which must be urged against reversing the headwaters of this stream. Reversal by tilting seems highly improbable; but more could be said for reversal by capture, on the ground that diversion of the long, roundabout Passaic-Millstone-Stony Brook-Assanpink-Delaware to a short course * to the sea via the Raritan would be promoted by the presence of a resistant barrier, the Rocky Hill trap ridge, athwart the longer river. After capture it is conceivable that a stream growing headward (southwestward) some twenty miles along the deserted valley of the beheaded river might develop into the present lower Millstone, although it seems problematical whether such a stream could push the divide beyond the resistant Rocky Hill barrier.

The fact that the Millstone gap through Rocky Hill is very narrow constitutes a more serious objection to the idea of prolonging the superposed subsequent of the Watchung area southwestward along the lower Millstone and so to the Delaware. In the field the contrast in size between Millstone gap on one hand, and the Paterson and Millburn gaps on the other, is easily observed. On maps the contrast is more fully appreciated, especially when we take into consideration the level to which the

---

* Approximately 10 miles as compared with 90 miles via the Delaware, on the basis of present river lengths. This may perhaps be accepted as some indication of the probable *degree* of contrast under other conditions of land and sea level, without denying the possibility that in the past conditions may have been less favorable to capture.

Millstone gap should be reduced. It would lie some 30 miles down stream from the Millburn gap; and assuming that there has been no great permanent warping * of the land since reversal of the Millstone, and that 113 feet above sea level is an approximate figure for the rock floor of Millburn gap (exclusive of the inner gorge of probable later date) it would seem that 100 feet or less should be the elevation of the rock floor in Millstone gap. But at this elevation the width of the latter gap is less than half a mile, as compared with widths of 1½ to 2 miles in the Paterson and Millburn gaps. The hard rock barrier itself is broader at the Millstone gap than at either of the others, and structural conditions may favor a narrower water gap in the former locality; whereas the larger volume of the supposedly continuous river farther down its course would favor a broader gap through Rocky Hill. It is not easy to balance these factors with any assurance, but it seems that in any case the degree of narrowness in the Millstone gap is so very marked as to negative the idea that a large river flowed through it in recent times.

There remain the possibilities (1) that the large river formerly passed through Rocky Hill, but was diverted at an earlier period than that in which the present rock floors of the Paterson and Millburn gaps were cut; or (2) that the floors of the gaps in all three areas are of approximately the same date, but warping has raised the Millstone gap above the level of its neighbors. In either case we should then look for a broad rock floor in the Millstone gap above the narrow gorge at the 100-foot level previously described. Now the upper part of this gap does widen gradually above the 100-foot level, the walls flaring out until the 150-foot or 200-foot level is reached, after which they again become more precipitous. But there is no nearly level floor such as occurs in the other gaps, and even at the 200-foot level the gap has a minimum width of less than a mile. The conditions suggest rejuvenation of the present Millstone, rather than the former presence of a larger stream across Rocky

* See earlier discussion of temporary warping during glacial period.

Hill. We therefore incline to the view that from the beginning the large subsequent stream of the Watchung area found its way to the sea along or near the present course of the Raritan (Fig. 21).

Larger Significance of the Superposed Drainage

The superposed drainage of the Watchung region described in the preceding pages implies a considerable thickness of beds in which there could be formed a very broad lowland and a prominent cuesta, drained by a large subsequent river and by obsequent streams some miles in length. The ancient topography was comparable to that of the present inner lowland and cuesta of the coastal plain in southern New Jersey, developed on hundreds of feet of slightly dipping Cretaceous strata. Apparently we can best secure the required thickness of beds by projecting inland the Cretaceous coastal plain. A thin deposit laid down during a minor oscillation or sub-cycle of post-Schooley date seems less likely to provide the conditions requisite for the drainage system which has left such impressive traces in northern New Jersey. Thus the nature of the superposed drainage, while perhaps too theoretical a matter to be classed as proof, may fairly be said to offer a certain measure of support to the theory that the streams had their inception on a thick coastal plain cover resting upon a peneplane of pre-Schooley age. But far more compelling reasons for entertaining this theory have been set forth in the earlier pages of this volume.

Derivation of the present drainage. — Apparently the present drainage may reasonably be derived from that described above through normal captures or early glacial displacements of the upper part of the superposed subsequent, as a result of which the waters were diverted from the Sparkill and Paterson gaps to more direct and less difficult pathways to the sea on weak Triassic sandstones and shales; by less conspicuous readjustments of streams to structure along other portions of the

Watchung ridges; and by later ice advances which shifted many streams to new and sometimes abnormal courses, blocked the Millburn gap with terminal moraine deposits, turned the drainage of the crescent back through the Paterson gap to form the present Passaic River, and simultaneously deflected drainage eastward from the east side of Millburn gap along the present path of the Rahway, approximately parallel to and just inside the arcuate terminal moraine.

The history of New Jersey drainage outlined in the preceding pages has formed the subject of repeated discussions with graduate students and research assistants in physiography during the past year. Out of these discussions have come many helpful suggestions for which acknowledgments are gladly made. The interpretation here offered seems to us to accord with a wider range of observed phenomena than does that suggested by Davis forty years ago, at a time when some of the pertinent facts had not yet been discovered. Whether the new interpretation can commend itself as favorably to the reader remains to be seen. In any case it seems worthy of critical examination, since it harmonizes most effectively with a theory of geomorphic evolution for which support is found in a wide variety of phenomena distributed over an extended area of the northern and central Appalachians.

REFERENCES

[1] Rollin D. Salisbury: "Passaic Folio," *U. S. Geol. Surv., Geologic Atlas*, Folio 157, 1–27, 1908, see p. 19.

[2] William Morris Davis and J. Walter Wood, Jr.: "The Geographic Development of Northern New Jersey," *Bost. Soc. Nat. Hist., Proc.*, 24, 365–423, 1889, see p. 407.

[3] Charles P. Berkey: Unpublished manuscript map kindly placed at our disposal.

[4] Rollin D. Salisbury and Henry B. Kümmel: "Lake Passaic, an Extinct Glacial Lake," *Geol. Surv. of New Jersey, Ann. Rep.*, for 1893, 225–328, 1894, see pp. 301–303.

[5] *Ibid.*, see pp. 319–325.

[6] Rollin D. Salisbury: "The Physical Geography of New Jersey," *Geol. Surv. New Jersey*, Vol. 4, 170 pp., 1898, see pp. 109–112.

CHAPTER EIGHT

CONCLUSION

The foregoing observations bring to its close a discussion which the reader has, I fear, found overlong and at times perhaps unduly intricate. If apology is needed, it may perhaps be found in the importance of a subject which not only touches the geomorphology of a large section of the Appalachians, but also, as will later appear, has indirect bearing on some fundamental problems of continental evolution. We have in these pages presented a new working hypothesis of Appalachian geomorphic evolution and have tested its applicability to two large regions of somewhat different physiographic aspect. While no attempt has been made to examine critically every phase of the theory of regional superposition of Appalachian streams from a pre-Schooley surface, or to weigh all its implications against those deriving from other hypotheses, enough has been written to show with sufficient fullness the general scope of the theory, the kind of considerations which seem to support it, and the nature of the consequences which would result from its acceptance. As we have seen, much may be urged in favor of the conception that southeast-flowing master streams of the Appalachian slope inherited their courses from a coastal plain cover which reposed upon a peneplane surface of pre-Schooley age and which formerly extended, in some places at least, from 125 to 200 miles northwest of the present inner margin of the coastal plain deposits. The necessary consequences of this conception, while to some degree opposed by generally accepted views on Appalachian and Atlantic coastal plain history, are in no case so definitely negatived by known facts as to justify us in discarding the theory of regional superposition. On the contrary, many of the deduced consequences harmonize so well with observed facts that one seems to find in the theory a simple and rational explanation for elements of Appalachian topography hitherto either inexplicable or to be explained only by an involved and complicated sequence of events.

In conclusion I should like to quote (with necessary changes in verb tense) two sentences with which Professor Davis prefaces his detailed analyses of Pennsylvania drainage, appropriating them as best expressing the attitude with which the present theory of Appalachian evolution is offered for criticism:

If the postulates that I use seem unsound and the arguments seem overdrawn, error may at least be avoided by not holding fast to the conclusions that are presented, for they are presented only tentatively. I do not feel by any means absolutely persuaded of the correctness of the results, but at the same time deem them worth giving out for discussion.

Such discussion will be all the more welcome because the primary object of this study has not been simply to discover the particular sequence of events in Appalachian drainage evolution. The ultimate goal at which we are aiming is the establishment of reliable criteria for determining the nature and amount of past changes in level of land and sea, as a means of discovering the causes and character of continental evolution. If the present study helps even a little in advancing our knowledge of Appalachian geomorphic history, it will contribute something toward the solution of one of the most fascinating major problems of physical geology.

INDEX

INDEX

The letters il following a page number indicate an illustration.

Acknowledgments, xiv, xv
Allegheny Front, 19 il
Ambrose's Brook, 81 il, 97, 99, 100
Anthracite River, 61 il, 63, 68, 69; upper, 70
Appalachians, vii, 5, 12, 15 il, 27, 31, 36, 37, 52, 56, 132; central, 29 il, 131; drainage, viii, xiv, 4, 10, 32, 36, 47, 48, 49, 52, 53, 56, 76, 86, 132, 133; evolution, ix, x, xi, 5, 12, 25, 43, 47, 52, 89, 132, 133; folds, viii, 14, 33, 34, 41, 55, 57, 58, 59, 60, 70; geomorphic history, 44, 47, 132, 133; geomorphology, xiii, 3, 76; highlands, viii; history, xiv, 5, 21, 56, 132; newer, 19 il; northern, 3, 5, 29 il, 47, 131; older, 19 il; oldland, 14; peneplane, 34; plateau, 19 il, 75; province, 35, 63, 89; region, viii, 21, 28, 33, 34, 47; ridges, 3, 98; slope, 132; southern, viii, 9; structures, 21, 33, 35, 37, 72; studies, 68; topography, xiv, 3, 132; water gaps, 12, 43, 44; wind gaps, 12, 43, 44
Appalachians, Pennsylvania, vii, 12
Ashley, G. H., 10
Assanpink Creek, 81 il, 128; lower, 128; upper, 81 il, 97, 98
Atlantic: border, 6, 33, 49; coastal plain, ix, x, xiii, 6, 9, 47, 48, 50, 60; coastal plain deposits, 48, 51; coastal plain history, 132, 133; ocean, xi; slope streams, 33

Baltimore, Md., 11, 39
Barrell, J., 4, 5, 6, 7, 8, 11, 26, 39, 43
Bascom, F., xv, 40
Beacon Hill, 86
Bear Mountain, N. Y., 36

Belvidere, N. J., 81 il
Bench: sub-alluvial, 38; wave-cut, viii
Berry Mountain, Pa., 36
Blauvelt, B., xiv
Blue Mountain, Pa., 4, 36, 37
Blue Ridge, Pa., 4
Bray, H., xiv
Broad Top Basin, 59, 73, 74
Brooklyn, N. Y., 81 il
Broome County, N. Y., 32
Butts, C., 50

Caldwell, N. Y., 89 il
Cambrian, 51; pre-, 40
Campbell, M. R., 21
Carboniferous: lowlands, 61 il
Center County, Pa., 73
Chamberlin, T. C., x
Chester Valley, Pa., 40
Chimney Rock Gap, N. J., 81 il, 125
Clark, W. B., 52
Coastal plain, 3, 4, 5, 8, 9, 10, 19 il, 22, 26, 27, 28, 31, 32, 35, 38, 40, 41 il, 47, 48, 49, 50, 51, 52, 67, 70, 71, 76, 77 il, 80, 83, 84, 86, 87, 89, 90, 91, 98, 99, 100, 101, 105 il, 109, 110, 130 (*see also* Atlantic); cover, ix, 4, 5, 9, 17 il, 22, 25, 28, 34, 35, 37, 38, 39, 43, 47, 49, 53, 58, 63, 66, 67, 68, 70, 71, 72, 73, 75, 77 il, 79, 84, 89, 91, 108, 109 (*see also* Cretaceous); cuestas, 93 il, 123 il; deposits, xiii, 4, 6, 11, 15 il, 21, 28, 38, 43, 47, 48, 50, 51, 61 il, 64 il, 81 il, 84, 85, 110, 117, 132; dissection, 98; floor, 38; Gulf, xi, 49, 50; history, 49, 133; New Jersey, 50, 52, 80, 98; paleontology, 48; pre-, 3, 6, 8, 9, 10, 11, 12, 14, 38; province, 14, 49; sediments, 25, 55, 80; series, 48,

49, 50, 80, 101; stratigraphy, 40, 48; Tertiary, 86; topography, 117; wedge, 9, 27, 31
Cols, 95
Columbia University, xiv, 91
Connecticut, 3, 25, 26, 28, 31, 32, 37, 53; central, 25; valley, 26 il, 27, 95
Connecticut River, 4, 28, 29 il, 31, 53; lower, 3, 25, 26, 26 il, 27, 28, 31, 33, 53; middle, 53; upper, 31
Continental: deposits, 49; evolution, 133; shelf, 49; shore, 50; theory of uplift, 8; warping, 50
Cooks Gap, Conn., 32
Cretaceous; age, 10, 14, 21, 29, 40, 43, 51, 52, 67, 76, 83, 84, 86; beds, 10, 14, 21, 27, 38, 39, 40, 53, 79, 86; coastal plain, 6, 26 il, 35, 67, 71, 73, 76, 84, 87, 130; deposits, 6, 15 il, 17 il, 39, 40, 47, 67, 76, 80, 84, 130; greensands, 51; late, 28, 58; marine transgression, 15 il, 28; peneplane, vii, 4, 14, 58; post-, 39, 67; sea, 27, 39, 52, 61 il, 65, 67, 79; series, 47, 79, 80; shoreline, 4; upper, 50
Cuesta, 91, 98, 99, 100, 101, 130; coastal plain, 93 il, 123 il; front, 101; scarp, 92
Cushetunk Mountain, N. J., 81 il

Darton, N. H., xv, 6, 8, 11, 39, 40, 52
Davis, W. M., xiii, xv, 3, 4, 5, 6, 8, 11, 21, 22, 25, 26, 26 il, 27, 35, 37, 39, 53, 55, 56, 57, 58, 59, 60, 63, 65, 66, 67, 68, 69, 70, 71, 72, 73, 74, 76, 79, 80, 83, 84, 85, 86, 89, 90, 96, 97, 98, 99, 100, 103, 105 il, 107, 108, 114, 115, 116, 127, 131, 133
Deerfield River, 29 il, 31
Delaware Bay, 98
Delaware River, 32, 61 il, 64 il, 81 il, 87, 89, 90, 101, 109, 127, 128; lower, 29 il, 33, 64 il; upper, 32
Delaware Water Gap, 81 il, End Papers
Deposits: alluvial fan, 50 (*see also* Coastal plain); continental, 49 (*see also* Cretaceous); delta, 48, 50; estuarine, 43, 61 il, 65, 67, 68; flood plain, 61 il, 66, 67; marine, 49

Derby, Conn., 31
Drainage: abnormal, 27, 91, 97, 131; adjustments, 21, 74, 79, 80, 83, 84, 85, 86, 87, 88, 89, 90, 91, 92, 95, 121, 130; antecedent, xiii, 33, 37, 57, 65, 79, 97, 98 (*see also* Appalachian); beheaded, 32, 36, 83, 87, 118, 121, 128; capture, 28, 33, 34, 35, 36, 43, 60, 63, 64, 65, 66, 68, 70, 71, 83, 88, 92, 95, 96, 101, 107, 114, 115, 116, 118, 121, 128, 130; complex, 72; consequent, vii, ix, xiii, 21, 25, 27, 28, 31, 34, 37, 53, 55, 57, 59, 60, 63, 65, 66, 67, 69, 70, 71, 72, 79, 91, 96, 97, 98, 101, 104, 108, 117, 118, 122, 125; consequent, initial, 28, 53, 55, 64, 90, 95, 96; diverted, 27, 32, 36, 64, 71, 91, 107, 110, 118, 121, 122, 128, 129, 132; evolution, xiv, 55, 56, 59, 133; history, 53, 55, 56, 57, 58, 59, 67, 70, 72, 79, 84, 97, 99, 101, 107, 108, 109, 114, 115, 116, 121, 126; incised, 21, 66, 83, 92, 95, 108; intrenched, 21, 83, 88, 91, 92, 101, 110; major, 60, 63, 89, 91, 103, 114, 117; master, viii, 27, 33, 34, 57, 60, 63, 65, 66, 67, 71, 87, 96, 108, 122, 125, 127, 132; New England, 31, 32; New Jersey, history, 84, 89, 90, 103, 131; obsequent, vii, 91, 92, 98, 100, 101, 103, 118, 122, 125, 130; pattern, 32, 35, 36, 55, 71, 90, 91, 95, 96, 100, 107, 128; Permian, 55; piracy, 63, 68, 96, 115, 116; plateau, 70, 75; rectilinear, 31, 33; rejuvenated, 21, 129; resequent, vii, 73, 91, 92, 99, 117, 118, 122, 125; reversed, 57, 59, 60, 63, 68, 97, 100, 101, 121, 127, 128, 129; subsequent, vii, ix, 28, 34, 35, 37, 53, 55, 57, 59, 64, 67, 68, 70, 83, 87, 91, 92, 95, 101, 103, 108, 109, 110, 111, 113, 114, 115, 116, 117, 118, 121, 122, 125, 126, 127, 128, 130; superposed, ix, 3, 4, 5, 21, 25, 28, 31, 35, 37, 43, 49, 55, 57, 58, 59, 63, 64, 65, 66, 67, 68, 70, 74, 80, 89, 90, 91, 92, 96, 100, 101, 103, 107, 108, 110, 111, 113, 114, 116, 117, 118, 121, 122, 125, 126, 127, 128, 130; transverse, 3, 74, 80, 83, 89, 90, 100, 104, 114, 116, 118, 121; trenched, 74, 87, 91, 92, 98. (*See also* Pennsylvania, Watchung)
Dutton, C. E., 7

INDEX

Easton, Pa., 32, 81 il
Elizabeth, N. J., 81 il, 107 il
Eocene, lower, 5
Exogyra, 48

Fall line, ix, 10, 39, 52, 54, 99; angle, 9, 101; barrier, 100; escarpment, 97, 100
Fall Zone, 26, 38, 52; epoch, 27; lowland, 49; peneplane 14, 21, 22, 26, 35, 37, 38, 39, 43, 47, 48, 49, 59, 73, 80, 84, 85, 88, 92, 99
Farmington River, 29 il, 31; upper, 31
Fault line, 53, 99; scarp, 53, 99
First Mountain, N. J., 36
Flemington, N. J., 81 il
Florida, viii
Franklin Furnace, N. J., 81 il
Frankstown Branch, 71, 72 (*see also* Juniata River)

Gaps: aligned, 37, 69, 104; alignment of, 37, 69, 107, 110; oblique, 92; offset, 92, 95; paired, 90, 92, 104, 109 il, 126; water, vii, ix, 12, 36, 37, 43, 44, 59, 67, 68, 72, 79, 90, 95, 96, 103, 104, 107, 108, 110, 112, 113, 114, 115, 116, 117, 118, 121, 122, 125, 126, 127, 128, 129, 130, 131; wind, ix, 12, 32, 36, 37, 43, 44, 67, 95, 96, 103, 104, 107, 108, 109, 110, 112, 113, 114, 117, 118, 121, 122, 125, 126, 127, 128, 129, 131
Georgia, 37
German Valley, N. J., 81 il
Grand Canyon district, 7
Green Pond, N. J., 81 il
Greenwood Lake, N. J., 81 il
Gilbert, G. K., x, 90
Glauconite, 50, 51, 52
Glauconitic sands, 51
Great Notch, N. J., 81 il, 125
Great Valley, 19 il
Greensands: ancient, 51; Cretaceous, 51; modern, 51
Gulf coastal plain, xi, 49, 50

Hackensack, N. J., 81 il; River, 81 il
Hackettstown, N. J., 81 il
Hancock, W. Va., 32

Harrisburg, Pa., vii, 28, 36, 57, 58, 61 il, 63, 64 il, 65, 69
Harrisburg peneplane, 19 il, 21, 98
Haverstraw, N. Y., 81 il
Highlands, crystalline, 4, 28, 58, 76, 79, 88, 89, 116
Hobbs, W. H., 33
Hook Mountain, N. J., 81 il, 104, 108, 115, 116, 122
Hopatcong, Lake, N. J., 81 il
Hornerstown epoch, 51
Housatonic River, 4, 29 il; lower, 3, 31, 53; middle, 31; upper, 53
Hubbell, M., 31
Hudson River, 31, 81 il, 109, 111, 114, 123 il

Interpretation, two-peneplane, 8; reasons for accepting, 9

Johnson, D., viii, ix, x, xi
Jonas, A. I., 40
Juniata River, 36, 57, 59, 71, 72, 73, 74; Frankstown Branch, 71, 72; lower, 32
Jura, 14, 76; Cretaceous, 14, 26, 27; drainage, 27, 28; folding, 25
Jurassic, 14, 21; early, 27; late, 21

Kingston, N. Y., 101
Kittatinny: cycle, 21; peneplane, 4, 11, 12, 14, 21; pre-, xiii, 4, 25, 47, 72; question, 14; surface, 3, 8
Kittatinny Mountain ridge, 4, 81 il, 98
Knapp, Mrs. W., xv
Knopf, E. B., xv, 40
Kümmel, H. B., xv, 26, 86, 107, 111, 112, 113

Lake Hopatcong, N. J., 81 il
Lane, A. C., xv
Lehigh River, 29 il, 32, 36, 61 il, 64 il, 69 il
Lesley, J. P., vii
Lewis, J. V., 83
Little Schuylkill River, 37
Livingston, N. J., 105 il, 115
Livingston Sand and Gravel Company, 115
Lobeck, A. K., xv

Lockatong Creek, 81 il, 89
Long Branch, N. J., 81 il
Long Hill, Pa., 81 il, 115, 122
Lutherville, Md., 39

Mahantango Mountain, Pa., 36
Mahoning Hills, Pa., 36
Manalapan Creek, 81 il, 97, 98
Manhattan, N. Y., 81 il
Mansfield, G. R., 50, 51
Marine: agencies, 6; deposits, 4; erosion, 6, 7; forms, 7; invasions, 4, 6, 22, 28, 31, 43, 47, 49, 50, 60, 86; non-, 7; origin, 7; planation, 4, 5; terrace, 6
Martin, G. C., 52
Maryland, 61 il
Maryland Survey, 40
Massachusetts, 32; northwestern, 4
Mattabesset River, 32
Mauch Chunk Mountain, Pa., 37
Medina, 61 il; barrier, 63; beds, 73; ridge, 66
Mendham, N. J., 81 il, 89
Merrimac River, 29 il
Meyerhoff, H. A., 31
Middletown, Conn., 25, 27, 31
Millburn, N. J., 81 il, 103, 104, 109, 122, 126, 127; Gap, 104, 107, 107 il, 110, 112, 115, 116, 117, 122, 123 il, 125, 126, 128, 129, 131
Millstone Gap, N. J., 104, 107, 128, 129
Millstone River, 81 il, 97, 98, 99, 100, 101, 127, 128, 129; lower, 99, 100, 103, 128; upper, 89 il, 103
Miocene, 86
Mississippi embayment, 9, 37
Montville, N. J., 105 il, 116
Morristown, N. J., 81 il
Mount Desert Island, 7
Mugler, D., xv
Murray, J., 51
Musconetcong River, 77 il, 80, 81 il; valley, 76

Navesink: epoch, 51; highlands, 81 il; River, 81 il
Nesquehoning Mountain, Pa., 37
Nevada-land, 7
Newark, N. J., 81 il

Newark: beds, 14, 15 il; deposition, 60; depression, 33; post-, 15 il
New Britain, Conn., 32
New Brunswick, N. J., 81 il, 100
New England, 4, 6, 8, 11, 53; central, 14; drainage, 31, 32; peneplane, 11; southern, 6, 7, 26; terraces, 7, 8; upland, 6, 11
New Jersey, 4, 12, 50, 51, 61 il, 76, 80, 84, 108; coastal plain, 50, 52, 80, 98; drainage, 3, 55, 58, 76, 84, 89, 90, 97, 103, 131; northern, xiv, 28, 30, 76, 77 il, 80, 81 il, 85, 130; southern, 130; topographic maps, 98; topographic survey, 103
New York, 4, 31, 32, 33, 61 il, 69
Nittany: anticlinal arch, 73, 74, 75; district, 73
North America, 50

Oriskany, 71

Paleozoic: beds, 40, 80, folded, 35; sediments, 37, 41 il, 58; stratigraphy, 50; time, 50
Palisades, 81 il, 109, 110, 123 il; trap ridge, 77 il, 85, 109
Passaic, N. J., 81 il, 104
Passaic River, 81 il, 89, 105 il, 111, 128, 131; upper, 114, 115
Paterson, N. J., 81 il, 103, 104, 107, 109, 111, 122, 125, 126, 127
Paterson Gaps, 104, 107, 107 il, 108, 109, 110, 112, 114, 115, 122, 123 il, 126, 128, 129, 130, 131
Paulins Kill River, 81 il
Peapack, N. J., 81 il
Pediment, rock, 38
Pen Argyl, 32
Pennsylvania, vii, xi, 4, 12, 28, 32, 37, 50, 55, 61 il, 72, 80, 81 il, 84; central, 4; drainage, 55, 56, 57, 59, 60, 69, 72, 73, 76, 89, 100, 133; eastern, xiv, 56, 63; features, 56; ridges, 14
Pequannock River, 81 il, 89, 103, 104, 105 il
Pequest River, 81 il
Permian rivers, 55; time, 58, 73
Philadelphia, Pa., 40, 61 il, 64 il

INDEX

141

Philadelphia Folio, 40
Piedmont, 6, 7, 8, 11, 19 il; terraces, 7, 8; upland, 6, 11
Pisgah Mountain, Pa., 37
Planation: fluvial, 5; marine, 4; process of, 27
Pleistocene, 39, 40
Plymouth Valley, Pa., 40
Pocono, 69 il, 73; plateau, 81 il; ridges, 37, 66; sandstone, 63, 65; synclines, 69
Pohatcong River, 81 il
Pompton River Gap, 81 il, 108
Port Jervis, N. Y., 32
Port Mercer, N. J., 81 il, 128
Potomac River, 29 il, 32, 33
Pottsville: beds, 73; ridges, 37
Powell, J. W., vii
Preuss, C., xiv
Princeton, N. J., 81 il, 100. 101, 103, 127
Profile: projected, xiv, 6, 10, 11, 12, 26, 43, 53; study, 100

Raccoon Ridge, Pa., 36
Rahway River, 81 il, 89, 104, 105 il, 114, 131
Raisz, E. J., xiv, xv
Ramapo Mountains, Pa., 81 il
Ramapo River, 81 il, 108
Raritan Bay, 81 il, 98; ancient, 64 il; North Branch, 81 il, 89; River, 29 il, 81 il, 101, 123 il, 127, 128, 130; South Branch, 81 il
Reading, 61 il, 64 il; Prong, 19 il
Renard, A. F., 51
Renner, G. T., 8, 52
Rhode Island, 37
Riker Hill, N. J., 81 il, 104, 105 il, 107 il, 108, 110, 112, 114, 115, 116, 122, 123 il
Rockaway River, 81 il, 89, 104, 105 il, 114
Rocky Hill, 81 il, 97, 128, 129, 130; barrier, 128; trap ridge, 97, 99, 100, 128
Rogers, W. B., vii
Rom, C., xiv, xv, 110

Saco River, 29 il
Salisbury, R. D., 86, 107, 107 il, 111, 127
Sandy Hook, N. J., 81 il
Saprolite, 34
Schooley: baselevel, 91, 92, 95; cycle, 21, 27, 28, 34, 35, 36, 43, 57, 59, 64, 66, 67, 68, 71, 72, 73, 74, 80, 85, 88, 89, 91, 110; denudation, 43; erosion, 39; events, 44; peneplanation, 27; peneplane, xiv, 4, 7, 11, 12, 14, 17 il, 19 il, 21, 22, 26, 33, 35, 36, 38, 41 il, 43, 58, 67, 68, 77 il, 79, 80, 84, 85, 86, 88, 93 il, 95, 105 il; surface, 3, 5, 8, 11, 12, 22, 33, 36, 39, 43, 58, 59, 65, 67, 68, 84, 101; pre-cycle, 53, 87, 89, 92, 93 il, 123 il, 132; pre-peneplane, xiii, 4, 25, 47, 55, 72, 77 il, 130; pre-surface, xiiii, 11, 55, 80, 87, 132; post-, 22, 44; post-erosion, 39, 101; post-peneplane, 11, 97; post-time, 4, 36, 53, 67, 80, 85, 88, 95, 110, 130; question, 14
Schuchert, C., 50
Schuylkill River, 29 il, 32, 36, 37, 57, 61 il, 63, 64 il; ancient, 64 il
Schwartz, M., 114
Scranton, Pa., 61 il, 64 il, 69
Sea, epicontinental, 47, 49, 50
Second Mountain, Pa., 36, 37
Sharp Mountain, Pa., 37
Shaw, E. W., 6, 8, 10
Shetucket River, 29 il, 31
Shickshinny Creek, 37
Shields, H., xiv
Siluro-Devonian, 65
Somerville, N. J., 81 il
Somerville peneplane, 19 il, 21
Sourland Mountain, N. Y., 81 il
South River, N. J., 81 il
Sparkill Gap, N. J., 81 il, 109, 110, 123 il, 130
Spring Creek, 74
Staten Island, N. Y., 81 il
Stephenson, L. W., xv, 48, 49, 50, 51
Stony Brook, 81 il, 101, 103, 127, 128
Suffern, N. J., 81 il
Summit, N. J., 103
Superposition, 4, 26 il, 31, 32, 33, 34, 36, 37, 43, 48, 55, 57, 59, 60, 65, 66, 69, 71, 72, 83, 85, 89, 92, 95, 105 il, 110, 114; local, 57, 67, 68, 69; process of, ix; regional, xiv, 17 il, 34, 66, 68, 69, 74, 75, 80, 83; superimposition, 80; theory of, 55, 57, 58, 59, 63, 64, 66, 89, 103; theory of regional, xiii, xiv, 10, 17 il, 29 il, 34,

36, 37, 47, 48, 51, 52, 53, 55, 64 il, 66, 69, 70, 72, 76, 77 il, 89, 90, 91, 92, 132, 133

Susquehanna River, vii, 33, 36, 57, 58, 61 il, 63, 64 il, 65, 66, 67, 68, 69, 87, 109; ancient, 61 il, 64 il; embryonic, 63, 65, 66, 68; lower, 3, 32, 33, 39 il; middle, 32; North Branch, 29 il, 32, 37, 61 il, 64 il; present, 61 il; upper North Branch, 69, 70; West Branch, 61 il, 64 il, 70, 74

Tarr, R. S., 3, 4, 25, 26 il, 27, 53
Tarrytown topographic quadrangle, 109
Tennessee, viii
Tennessee River, viii
Terraces, 6, 7, 8; local, 10; marine, 6; monoclinal, 7; New England, 7; Piedmont, 7; slope, 6, 10; wave-cut, 6
Tertiary: age, 21, 27, 52, 58, 86, 97; beds, 86; coastal plain, 86; early 28, 47, 58, 128; late, 39, 86; seas, 52; series, 47
Theory of: antecedence, 37; Appalachian evolution, 3, 5, 25, 44, 47, 52; continental uplift, 8; fracture control, 33; geomorphic evolution, 13; reversal, 60; stream adjustments, 59, 66, 69, 70, 72 (*See also* Superposition, theory of regional)
Trenton, N. J., 61 il, 64 il, 81 il
Trias-Jura-Cretaceous, 14
Triassic, 29 il, 103; basins, 25; belts, 27, 28, 29; formations, 53, 79, 84, 85, 89, 93 il, 99, 101, 103, 110, 115; fragments, 79; Lowland, 19 il, 31, 53, 77 il, 123 il; pre-, 35, 77 il; regions, 79, 84, 85, 89; sandstones, 25, 100, 127, 130; shales, 25, 83, 130; traps, 127
Tuscarora barrier, 63
Tuscarora Mountain, Pa., 36

Tussey's Mountain, Pa., 71, 72
Twenhofel, W. H., xv
Tyrone, Pa., 71

Ulrich, E. O., 50
Unionidae, viii
United States, xiii, xiv; topographic quadrangles, 103

Vermont, 4
Ver Steeg, K., 12, 43, 44, 68

Wallace, D., xv
Wallkill River, 81 il
Warrior Ridge, Pa., 71, 72
Washington, D. C., 32
Watchung Mountains, 81 il, 85, 87, 88, 103, 110, 123 il; Basin, 87; crescent, 90, 95, 103, 108, 115, 117, 122; drainage, 85, 86, 89, 105 il; region, 79, 84, 86, 87, 88, 96, 107 il, 109, 110, 113, 114, 126, 127, 128, 129, 130; ridges, 77, 81, 83, 84, 85, 86, 89, 104, 108, 110, 117, 121, 131; trap sheet, 87, 91, 93 il, 95, First, 79, 87, 92, 109, 112, 114, 118, 122, 125, 126, gap. 111, 113; Second, 79, 83, 89, 91, 92, 114, 118, 122, 125, 126, gap. 111, 113
Well-borings, 37, 38
Westfield River, 29 il, 31
Williamsport, Conn., 61 il, 64 il, 74
Willimantic River, 29 il, 31
Willis, B., xv, 8, 12, 34
Wisconsin time, 113
Wood, Jr., J. W., 115
Wyoming: basin, 37; syncline, 69

Yonkers, N. Y. 81 il

Zernitz, E., xiv

COLUMBIA UNIVERSITY PRESS
COLUMBIA UNIVERSITY
NEW YORK

FOREIGN AGENT
OXFORD UNIVERSITY PRESS
HUMPHREY MILFORD
AMEN HOUSE, LONDON, E.C.

Bei Fragen zur Produktsicherheit wenden Sie sich bitte an:
If you have any questions regarding product safety,
please contact:

Walter de Gruyter GmbH
Genthiner Straße 13
10785 Berlin
productsafety@degruyterbrill.com